SIGNAL RECOVERY TECHNIQUES FOR IMAGE AND VIDEO
COMPRESSION AND TRANSMISSION

SIGNAL RECOVERY TECHNIQUES FOR IMAGE AND VIDEO COMPRESSION AND TRANSMISSION

by

AGGELOS KATSAGGELOS
Northwestern University

and

NICK GALATSANOS
Illinois Institute of Technology

KLUWER ACADEMIC PUBLISHERS
BOSTON / DORDRECHT / LONDON

A C.I.P. Catalogue record for this book is available from the Library of Congress.

ISBN 978-1-4419-5063-5

Published by Kluwer Academic Publishers,
P.O. Box 17, 3300 AA Dordrecht, The Netherlands.

Sold and distributed in North, Central and South America
by Kluwer Academic Publishers,
101 Philip Drive, Norwell, MA 02061, U.S.A.

In all other countries, sold and distributed
by Kluwer Academic Publishers,
P.O. Box 322, 3300 AH Dordrecht, The Netherlands.

Printed on acid-free paper

Contents

7
ERROR CONCEALMENT IN ENCODED VIDEO STREAMS   199
Paul Salama, Ness B. Shroff, and Edward J. Delp

8
ERROR CONCEALMENT FOR MPEG-2 VIDEO   235
Susanna Aign

Preface

The fields of lossy image and video compression and transmission of compressed image and video experienced an explosive growth recently and have resulted in a number of very important technological advances. The field of signal recovery on the other hand is a relatively mature research field, to which many ideas from applied mathematics and estimation theory have been applied to.

Traditionally the fields of signal recovery and compression have developed separately because the problems they examine seem to be very different, and the techniques used appear unrelated. Recently though, there is a growing consent among the research community that the two fields are quite related. Indeed, in signal recovery it is attempted to reconstruct the best possible signal from limited information. Similarly, any decoder of a compressed image or video bitstream is faced exactly with the same problem. In addition, in every image and video transmission system information is lost due to channel errors. Therefore, signal recovery techniques need to be used in the receiver to reconstruct the best possible signal from the available information and thus "conceal" the channel errors. One of the objectives of this book is therefore to help establish "a bridge" between the signal recovery and image and video compression and transmission communities.

In lossy image and video compression the problem at hand is to represent the source information in a very compact way without altering much its visual quality. Expressed in a different way, the goal of image and video compression is to trade the rate in representing the original source with the resulting distortion. The generic steps towards this goal, are the mapping or transformation of the source data to a different domain, with the intention of decorrelating them, and then the quantization of the parameters of the resulting representation. For example, for still images, the original intensity data are block transformed and the resulting transform coefficients are quantized. Similarly, for video data, the temporal decorrelation is first achieved with the use of motion compensation, and the resulting compensation error is block transformed and quantized. The quantization step is the one responsible for the rate reduction, but also for the *intentional* introduction of distortion or artifacts.

During transmission of compressed images and video two types of errors are typically encountered: *random bit and erasure* errors [1]. The former usually corrupt random patterns of bits of the transmitted signal, while the latter result in the loss of continuous segments of bits of the transmitted signal. In both cases, since variable length codes are used to further compress the transmitted information, either of these errors can cause desynchronization of the transmitted bit-stream. In addition, because information is typically predictively coded, errors can propagate through a decoded sequence over time. These are the *unintentional* errors, which need to be avoided by appropriately designing the system layer. However, there will always be residual errors which need to be concealed at the decoder.

In signal recovery applications the objective is, as the name implies, to recover or reconstruct or restore information about the original signal which has been lost. In the traditional applications of signal recovery the information loss is due to the band-limited (usually low-pass) nature of most imaging systems, motion between the camera and the scene, atmospheric turbulence, and noise which in many cases corrupts the data. Since the available data can not uniquely characterize the underlying signal, prior knowledge about the signal is typically used in an effort to uniquely characterize it. For this purpose, techniques from estimation theory, such as, Bayesian Inference [2], and from areas of applied mathematics, such as, the theory of projections onto convex sets (POCS) [3] have been applied over the past 20 years to the signal recovery problem.

The first three chapters of the book deal with the recovery of information which is lost due to the quantization process in compressing still images and video using one of the available compression standards. It is demonstrated how well established recovery techniques, which had been already developed and applied to more traditional recovery problems, can be modified and extended to find application to the recovery of information lost due to the quantization process.

The first chapter by Molina, Katsaggelos and Mateos *Removal of Blocking Artifacts Using a Hierarchical Bayesian Approach* formulates the still image decoding problem using the hierarchical Bayesian estimation paradigm. Such formulation allows for the simultaneous estimation of both the decoded image and the hyperparameters which are used to model the prior information. Furthermore, it allows for the rigorous combination of the hyperparameters estimated at the encoder and the decoder.

The second chapter by Llados, Robertson and Stevenson *A Stochastic Technique for Removal of Artifacts in Compressed Images and Video* presents a maximum a posteriori (MAP) estimation approach to the problem of decoding image and video signals. This approach is based on a non-Gaussian Markov Random field image prior model. This model is used to introduce prior knowledge into the estimation process using also the Bayesian signal estimation paradigm. In this chapter the image decoding problem is formulated as a convex constrained minimization problem and efficient gradient-projection algorithms are explored to solve it. Numerical examples are shown with both JPEG and H.263 compressed images and video.

The third chapter by Yang, Choi and Galatsanos *Image Image Recovery from Compressed Video Using Multichannel Regularization* develops a deterministic recovery approach which is based on the well-known regularization principle. A multi-channel framework is applied to this recovery problem according to which both spatial and temporal prior knowledge is utilized to reconstruct the video signal from the transmitted data. One of the objectives of this chapter is to demonstrate the advantages of incorporating temporal prior knowledge into the solution approach, in addition to the spatial prior knowledge.

The last six chapters in the book deal with the recovery of information lost during the transmission process of compressed images and video. The sixth chapter deals with the transmission of images compressed by a subband coder, while the remaining chapters deal with the transmission of video compressed with the use of an existing standardized coder.

Unlike the rest of the chapters in which standardized codecs, such as, JPEG, MPEG, H.261 and H.263 are considered, the fourth chapter by Hemami and Gray *Subband Coded Image Reconstruction for Lossy Packet Networks* treats a subband based codec for still images. Recovery techniques are examined to ameliorate the errors that can occur when the subband coded image is transmitted over lossy packet networks. Interpolation techniques are investigated that exploit the image correlations both within and between subbands to reconstruct the missing information.

The fifth chapter by Banham and Brailean *Video Coding Standards: Error Resilience and Concealment* presents the tools provided by video coding standards for ameliorating the effects of transmitting standardized video bitstreams in error prone environments. More specifically in this context, both ITU-T standards, such as, H.261 and H.263, and ISO/IEC standards, such as, MPEG-1, MPEG-2, and MPEG-4 are discussed.

The sixth chapter by Girod and Farber *Error-Resilient Standard-Compliant Video Coding* explores the idea of a feedback channel between the coder and the decoder within the error resilience compressed video transmission context. More specifically, errorless transmission is assumed for this channel which carries acknowledgement information. The availability of this information, allows the coder to retransmit the lost information. Simulation experiments are presented using bursty error sequences and wireless channels at various signal to noise ratios.

The seventh chapter by Salama, Shroff and Delp *Error Concealment in Encoded Video Streams* develops recovery solutions to the compressed video transmission problem. More specifically, both spatial and temporal information in lost macroblocks is recovered using deterministic and stochastic approaches. For the deterministic approaches interpolation and minimization of inter-sample variation from its uncorrupted neighbors is used. For the stochastic approaches Bayesian estimation is applied once more using a Markov Random Field for the image prior. The goal of this chapter is to prevent the loss of ATM cells from affecting the video decoding process.

The eighth chapter by Aign *Error Concealment for MPEG-2 Video* discusses concealment techniques applicable to MPEG-2 compressed video. More specifically, filtering and interpolation techniques are used to recover both spatial and temporal information which has been lost due to transmission over error-prone channels. A number of experimental results demonstrare the effectiveness of various techniques.

The final chapter of this book by Ngan and Yap *Combined Source-Channel Video Coding* presents a complete error resilient H.263-based compressed video transmission system. This system uses channel-coding and error resilient entropy coding to combat the effects of the error-prone channel. More specifically, rate compatible punctured convolutional (RCPC) codes are used to provide unequal protection in a mobile channel environment.

One of the objectives of the chapters in the book (and the references therein) is to demonstrate the application of well established recovery techniques to new problems which arise when images and video are compressed and transmitted over error-prone channels. Another objective is to motivate the introduction of the decoder-estimator paradigm, according to which the decoder is not simply undoing the operations performed by the encoder, but instead solves an estimation problem based on the available bitstream and prior knowledge about the source image or video. This then could eventually lead to new types of encoders in which the information hardest to be recovered by the decoder (high frequency information?) is only transmitted, while the information which is easier to be recovered by the decoder (low frequency information?) is thrown away.

References

[1] Wang Y. and Zhu Q.-F. "Error Control and Concealment for Video Communication: A Review," *Proceedings of the IEEE*, Vol. 86, No. 5, pp.974-997, May 1998.

[2] Berger J. O., *Statistical Decision Theory and Bayesian Analysis,* Springer-Verlag, New York, 1985.

[3] Stark H., and Yang Y., *Vector Space Projections: A Numerical Approach to Signal and Image Processing, Neural Networks and Optics* John Wiley and Sons 1998.

Aggelos K. Katsaggelos
Nikolas P. Galatsanos
Chicago, Illinois

Acknowledgement

We would like to thank all the authors of the chapters for their excellent contributions to this book. Special thanks are due to Chun-Jen Tsai at Northwestern University who helped us with various LaTeX and postscript aspects. His expertise and willingness to help made this book a reality. Finally we want to thank Mike Casey and James Finlay at Kluwer. Mike was instrumental in starting this project and James in making sure the final manuscript reached his desk.

To my mother Ειρήνη and late father Κωνσταντίνος (A.K.K.)
and
to my mother Δέσποινα and late father Παναγιώτης (N.P.G.)

1 REMOVAL OF BLOCKING ARTIFACTS USING A HIERARCHICAL BAYESIAN APPROACH

Rafael Molina[1] Aggelos K. Katsaggelos[2] and Javier Mateos[1]

[1] Departamento de Ciencias de la Computación e I.A.,
Universidad de Granada,
18071 Granada, Spain.
{rms,jmd}@decsai.ugr.es

[2] Department of Electrical and Computer Engineering,
Northwestern University,
2145 Sheridan Road, Evanston, IL 60208-3118.
aggk@ece.nwu.edu

Abstract:
Blocking artifacts are exhibited by block-compressed still images and sequences of images, primarily at high compression ratios. These artifacts are due to the independent processing (quantization) of the block transformed values of the intensity or the displaced frame difference. In this chapter, we provide a survey of the literature on the removal of such artifacts. We also apply the hierarchical Bayesian paradigm to the reconstruction of block discrete cosine transform compressed images and the estimation of the required hyperparameters. The evidence analysis within the hierarchical Bayesian paradigm is used to derive expressions for the iterative evaluation of these parameters. The proposed method allows for the combination of parameters estimated at the coder and decoder. The performance of the proposed algorithms is demonstrated experimentally.

2

Acknowledgments

The work reported in section 1.7 was done in collaboration with B. Jímenez and C. Illia, at the University of Granada. This work has been partially supported by the "Comisión Nacional de Ciencia y Tecnología" under contract TIC97-0989

1.1 INTRODUCTION

Block-transform coding is by far the most popular approach for image compression. Evidence of this fact is that both the Joint Photograph Experts Group (JPEG) and the Motion Pictures Experts Group (MPEG) recommend the use of the block discrete cosine transform (BDCT) for compressing still and sequences of motion images, respectively [18, 51].

With such methods, an image is divided into small blocks, usually 8×8 pixels, which are transformed using the Discrete Cosine Transform (DCT). Note that BDCT is, by itself, lossless (except for rounding operation) and it can be reversed without decreasing the amount of information present in the image. The compression is mainly achieved by quantizing the DCT coefficients, that is, by trying to discard the information that is not visually relevant. Quantization is performed on a block basis using a quantization matrix whose elements represent a quantization interval for each DCT coefficient of the block. The entries of the quantization matrix are chosen depending on the visibility of each coefficient; the less important DCT coefficients, responsible for the very high frequencies, are more coarsely quantized than the low frequency coefficients. After quantization, the blocks are run-length encoded using a predetermined scan and the result is entropy coded.

A standard decoder just reverses the process. The entropy coded data are decoded, dequantized and the inverse BDCT is performed. The obtained image is not exactly the original one but, if the quantization coefficients are carefully chosen, the image will appear very similar to the original one.

Let us state more formally the reconstruction problem. Throughout this chapter a digital $M \times N$ image is treated as an $(M \times N) \times 1$ vector in the $R^{M \times N}$ space by lexicographically ordering either its rows or columns. The BDCT is a linear transformation from $R^{M \times N}$ to $R^{M \times N}$. Then, for an image $\mathbf{f}$ we can write

$$\mathbf{F} = \mathbf{Bf},$$

where $\mathbf{F}$ is the BDCT of $\mathbf{f}$ and $\mathbf{B}$ is the BDCT matrix. To achieve a bit-rate reduction, each element of $\mathbf{F}$ is quantized. This quantization operator represents mathematically a mapping, $\mathcal{Q}$, from $R^{M \times N}$ to $R^{M \times N}$. The input-output relation of the coder can be modeled by

$$\mathbf{G} = \mathcal{Q}\mathbf{Bf}.$$

Due to the unitary property of the DCT matrices, the BDCT matrix is also unitary and the inverse transform can be simply expressed by $\mathbf{B}^t$, where t denotes the transpose of a matrix. In the receiver only the quantized BDCT coefficients, $\mathbf{G}$, are available and the output of a conventional decoder is given by

$$\mathbf{g} = \mathbf{B}^t Q^{-1} \mathbf{G}.$$

Such a compression method results in blocking artifacts if the DCT coefficients are coarsely quantized. These artifacts manifest themselves as artificial discontinuities between adjacent blocks. It is a direct result of the independent processing of the blocks which does not take into account the between-block pixel correlations. They constitute a serious bottleneck for many important visual communication applications that require visually pleasing images at very high compression ratios. *The reconstruction problem calls for finding an estimate of* $\mathbf{f}$ *given* $\mathbf{g}$, Q *and possibly knowledge about* $\mathbf{f}$. The advances in VLSI technology will result in the incorporation of recovery algorithms at the decoders, and will bridge the conflicting requirements of high-quality images and high compression ratios.

In this chapter we first review the literature on removing blocking artifacts, and then formulate, within the hierarchical Bayesian paradigm, a new algorithm which reconstructs the image and estimates the unknown hyperparameters (regularization parameters) at the same time. The regularization parameters can also be estimated at the coder using the original image, and after transmission they can be combined at the decoder with the ones obtained from the reconstructed image. We show how these combinations can be made, rigorously, within the hierarchical Bayesian approach to the reconstruction problem. The results are also extended to color images.

The rest of the chapter is organized as follows. In section 1.2 we survey the literature on removing blocking artifacts. In section 1.3 we describe the hierarchical Bayesian paradigm for the image reconstruction problem under consideration. Section 1.4 describes the prior and noise models for the reconstruction problem. In section 1.5 we describe the use of the hierarchical approach to the hyperparameter estimation problem and the recovery algorithm is presented for both the reconstructed and the original image. In section 1.6 we examine how the parameters obtained from the original and reconstructed images can be combined following the hierarchical Bayesian paradigm. The method is extended to color images in section 1.7. Experimental results are presented in section 1.8 and, finally, section 1.9 concludes the chapter.

1.2 A SURVEY OF METHODS TO REMOVE BLOCKING ARTIFACTS

In the past various algorithms have been proposed to improve the quality of block-transform compressed images at the decoder without increasing the bit-rate. We can distinguish between two different approaches. With the first one the original DCT coefficients are estimated at the decoder, so that they become similar to the ones the original image had and, thus, obtaining a reduction of the blocking artifacts and an improvement of the image visual quality. With the second approach the image is filtered in the spatial domain in order to mainly smooth the block boundaries and obtain an image more pleasing to the human visual system. The above two approaches do not modify the JPEG standard. Approaches which do, will not be discussed here (see for instance [8, 15, 65, 67, 68]).

It has been reported in the literature that the frequency domain reconstruction approaches are faster than the spatial domain ones. We review both approaches next.

1.2.1 DCT Domain Approaches

A technique for predicting the first five low frequency coefficients was recommended by the JPEG Group in the K annex of the "Requirements and Guidelines" document [18]. However, in areas with sharp intensity transitions such a prediction scheme fails. It therefore was necessary to find alternative methods. The most common approach to find an image with reduced blocking artifacts in the frequency domain is to estimate the DCT coefficient so that a visual measure of the blockiness is minimized [22, 23, 43].

Yang *et al.* [77] propose a Constrained Least Squares (CLS) regularized recovery approach. All DCT coefficients are estimated by trading image smoothness with fidelity to the received data. Although the problem is formulated in the spatial domain, the optimization is performed iteratively in the frequency domain. Crouse and Ramchandran [11] adapt the method proposed by Yang *et al.* [77] to incorporate information about the local characteristics by estimating the regularization parameter for each block. Following this line, Choy *et al.* [9] estimate the DCT coefficients using a weighted least squares approach. The weights are locally estimated using the local mean and variance.

Paek and Lee [49] use the Projection onto Convex Sets (POCS) theory to smooth the block boundaries in the DCT domain. Assuming that the original image is highly correlated, the global frequency characteristics in two adjacent blocks may be similar to the local ones in each block. Thus, they consider the high frequency components of the global characteristics that do not exist in the local ones as the effect caused by the blocking artifact. The parameters required by the POCS method are empirically estimated.

A curious way to reduce the blocking artifacts is proposed in [13, 70, 42]. The main idea is to introduce a distortion on the DCT coefficients. This distortion (usually using dithering) makes the blocking to 'disappear' but increases the mean square error and reduces the peak signal to noise ratio. The technique proposed by Webb [70] introduces this distortion only at high detailed zones, estimating the four low DCT coefficients on the smooth zones. The classification of smooth or detailed zones is performed with the use of chrominance information.

Silverstein and Klein [60] also propose to introduce a distortion on the DCT coefficients. The idea is to determine at the coder which blocks can be improved by post-processing and to flag them by adjusting a single DCT coefficient so that the sum of the coefficients of the block contains the flag for post-processing as the parity of the block. This distorts the image by a very tiny amount. The end result is a compressed image that can be decompressed on any standard JPEG decompressor, but which can be enhanced by a sophisticated decompressor.

Other proposed methods correct the DCT coefficients of a block by smoothing the pixels at the neighboring block borders so that the resultant image has minimum block boundary discontinuity (defined as the sum of the squared differences of pixel values along the block boundary) [21] or perform the estimation of the quantization matrix at the coder [52, 53, 54].

1.2.2 *Spatial Domain Approaches*

Spatial domain approaches were introduced in the mid eighties by Reeves and Lim [56], Ramamurthi and Gersho [55], Tzou [69], and Baskurt *et al.* [2], by applying filtering and restoration techniques to the blocking artifact removal problem. However, it has been in this decade when the reconstruction of block compressed images in the spatial domain has flourished. An enormous amount of papers has been published on compressed image recovering techniques using a great variety of methods in the spatial domain. Perhaps, the most influential works have been developed by Sauer [59], Zakhor [79] and Yang *et al.* [77, 78].

Sauer pointed out in [59] that: "Classical spatially-invariant filtering techniques are of little use in removing this signal-dependent reconstruction error". A bit later, Zakhor [79] presented a reconstruction algorithm, commented by Reeves and Eddins [57], based on the POCS theory, using two convex sets. One of the convex sets is equivalent to a convolution with an ideal low-pass filter and the other one deals with the quantization intervals of the transform coefficients. Conclusions were similar to Sauer's; "the low-pass filtering by itself could remove the blockiness but at the expense of increased blurriness". Yang *et al.* [77] used well grounded approaches to provide rigorous solutions to the problem of the removal of blocking artifacts. The first method is based on the theory of POCS while the second on the CLS approach. For the POCS-based method, a new constraint set is defined that conveys smoothness information not captured by the transmitted BDCT coefficients and the projection onto it is computed. For the CLS method an objective function is proposed which captures the smoothness properties of the original image. The POCS based method was later extended to include spatial adaptivity [78]. The POCS theory was also applied in [4] to the reconstruction of images compressed using vector quantization.

Taking into account the necessity of spatially adaptive techniques, different methods have been proposed. Stevenson [62] proposed a stochastic regularization technique using a statistical model for the image based on the convex Huber prior function. This function is quadratic for values smaller than a threshold, like block boundaries, and linear for greater values like natural discontinuities. Since the prior and penalty functions are convex, the MAP estimation can be performed by gradient descend. This initial work has been developed in [46, 47, 63] and was the basis for the works of Luo *et al.* [31, 32] where a Huber-Markov Random Field based model is optimized by POCS [31] or using the iterated conditional mode (ICM) technique [32]. They also incorporated two different values for the threshold; one for block boundaries and another for the pixels inside the block. Another Markov Random Field based method, using Mean Field Annealing for the MAP estimation, was proposed by Özcelik *et al.* [48]. This approach was applied to the removal of blocking artifacts in still images, as well as, video.

The method proposed by Yang *et al.* [77], and developed in [75] and [78], also had a great repercussion on the scientist community. Based on those works, Paek *et al.* [50] apply a similar technique to block boundaries and the pixels inside the block. Kwak and Haddad [26] optimize the method in [78] by canceling out the DCT-IDCT pair needed for the constraint on the quantization intervals of the transform coefficients.

In [39, 41], the method in [78] is formulated within the Bayesian hierarchical paradigm (see [45]), where the reconstruction and parameter estimation problems are tackled using well grounded techniques.

In a number of approaches, edge information is used for guiding the spatially adaptive filtering, which smooths block boundaries without altering the edges [25, 29, 30, 42, 73, 76]. Instead of using edge detection, some authors divide the image into regions with similar characteristics so then an adaptive filter is applied to each region depending on its local characteristics. Region classification is performed with the use of clustering [16], histogram and gradient information [24], local statistics [19, 20, 66], and other suitable techniques.

There are also a great number of empirical techniques which use one of various filters depending on the image local characteristics [12, 33, 58, 74]. Lai *et al.* [27, 28] adapt the method proposed in [79] for adaptive processing. They use just one iteration of the non-adaptive image filter proposed in [79] followed by DCT coefficient fitting based on interpolation of the neighboring blocks. After that, the method in [79] is iteratively applied to smooth blocks. This technique, as most of the previously described techniques, use certain parameters which are either chosen by the user or empirically estimated from the data.

We finally mention the existence of techniques based on the use of wavelets (see, for example, [17, 72], and references therein).

1.3 HIERARCHICAL BAYESIAN PARADIGM

The hierarchical Bayesian paradigm has been applied to many areas of research related to image analysis. Buntine [5] has applied this theory to the construction of classification trees and Spiegelhalter and Lauritzen [61] to the problem of refining probabilistic networks. Buntine [6] and Cooper and Herkovsits [10] have used the same framework for constructing such networks. MacKay [34] and Buntine and Weigund [7] use the full Bayesian framework in backpropagation networks and MacKay [35], following Gull [14], applies this framework to interpolation problems. The paradigm has also been applied to restoration problems [44, 45].

In the hierarchical approach to image reconstruction we have at least two stages. In the first stage, knowledge about the structural form of the noise and the structural behavior of the reconstruction is used in forming the conditional probability density functions $p(\mathbf{f}|\alpha)$ and $p(\mathbf{g}|\mathbf{f}, \beta)$, respectively. These noise and image models depend on the unknown hyperparameters or hypervectors α and β. In the second stage the hierarchical Bayesian paradigm defines a hyperprior on the hyperparameters, where information about these hyperparameters is included.

Although in some cases it would be possible to know, from previous experience, relations between the hyperparameters we shall study here the model where the global probability is defined as

$$p(\alpha, \beta, \mathbf{f}, \mathbf{g}) = p(\alpha)p(\beta)p(\mathbf{f}|\alpha)p(\mathbf{g}|\mathbf{f}, \beta). \tag{1.1}$$

With the so called *evidence analysis*, see [3, 36] for other possible names, $p(\alpha, \beta, \mathbf{f}, \mathbf{g})$ is integrated over $\mathbf{f}$ to give the likelihood $p(\alpha, \beta|\mathbf{g})$; this likelihood is then maximized over the hyperparameters. Recently, an alternative procedure has been suggested by Buntine and Weigund [7], Strauss, *et al.* [64], Wolpert [71], Molina [44] and commented by Archer and Titterington [1]. With this procedure, which is henceforth called the *MAP analysis*, one integrates $p(\alpha, \beta, \mathbf{f}, \mathbf{g})$ over α and β to obtain the true likelihood, and then maximizes the true posterior over $\mathbf{f}$. We show in [45] that the evidence analysis, the analysis we will follow here, is more appropriate than the MAP analysis for image restoration-reconstruction problems.

Let us now examine the components of the first and second stages of the hierarchical Bayesian paradigm for the deblocking problem under consideration.

1.4 IMAGE AND NOISE MODELS

Let us assume that $k \times k$ blocks are used to transform the $M \times N$ image, where M and N are multiples of k. Since for the removal of blocking artifacts we will only be operating on the block boundary pixels, we introduce the needed notation to characterize these image pixels.

Let $\mathbf{f}_{cl}$ be a column vector formed by stacking all the elements of $\mathbf{f}$ which are on the left of a block boundary column, cl, but not on a four-block intersection (see figure 1.1), that is,

$$\begin{aligned} \mathbf{f}_{cl} \quad = \quad & \{\mathbf{f}(u) \mid u = (x, y) \text{ with } x = k * i + l, \, y = k * j, \\ & i = 0, 1, 2, \ldots, M/k - 1, \, j = 1, 2, \ldots, N/k - 1, \, l = 2, 3, \ldots, k - 1\}. \end{aligned}$$

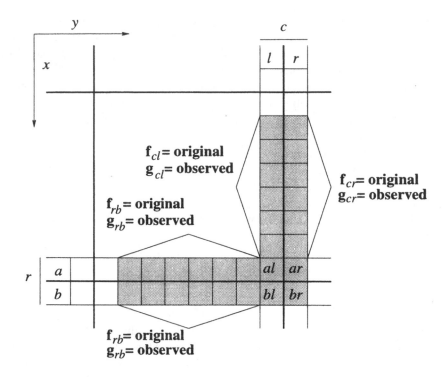

f_{cl} = original
g_{cl} = observed

f_{cr} = original
g_{cr} = observed

f_{rb} = original
g_{rb} = observed

f_{rb} = original
g_{rb} = observed

Figure 1.1   Distribution of boundary pixels according to their position.

Let also $\mathbf{f}_{cr}$ be the column vector formed by stacking all the elements of $\mathbf{f}$ that are on the right of a block boundary column, cr, but not on a four-block intersection (see figure 1.1). The same way, $\mathbf{g}_{cl}$ and $\mathbf{g}_{cr}$ are formed from $\mathbf{g}$.

We also define the vectors

$$\mathbf{f}_c^t = \{(\mathbf{f}_{cl}(i)\ \mathbf{f}_{cr}(i),\ i = 1, \ldots, p)\}, \qquad \mathbf{g}_c^t = \{(\mathbf{g}_{cl}(i)\ \mathbf{g}_{cr}(i),\ i = 1, \ldots, p)\},$$

with

$$p = (k - 2)(M/k) \times (N/k - 1). \tag{1.2}$$

Note that $\mathbf{f}_{cl}, \mathbf{f}_{cr}, \mathbf{g}_{cl}$ and $\mathbf{g}_{cr}$ are $p \times 1$ vectors.

In a similar way we define $\mathbf{f}_{ra}, \mathbf{g}_{ra}, \mathbf{f}_{rb}$ and $\mathbf{g}_{rb}$, the column vectors representing the rows above and below a row block boundary of $\mathbf{f}$ and $\mathbf{g}$, respectively (see figure 1.1). Related to them are the vectors

$$\mathbf{f}_r^t = \{(\mathbf{f}_{ra}(i)\ \mathbf{f}_{rb}(i),\ i = 1, \ldots, q)\}, \qquad \mathbf{g}_r^t = \{(\mathbf{g}_{ra}(i)\ \mathbf{g}_{rb}(i),\ i = 1, \ldots, q)\},$$

with

$$q = (k - 2)(N/k) \times (M/k - 1). \tag{1.3}$$

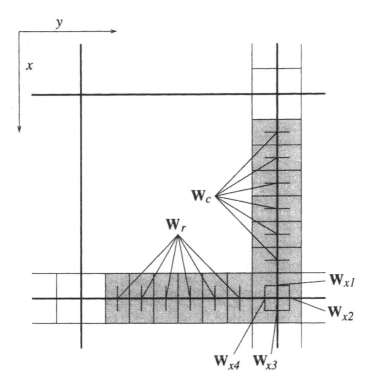

Figure 1.2   Distribution of weights.

We further stack all the elements of $\mathbf{f}$ above a horizontal boundary and to the left of a vertical boundary, indicated by al in figure 1.1, into vector $\mathbf{f}_{al}$, that is,

$$\mathbf{f}_{al} \;=\; \{\mathbf{f}(u) \mid u = (x, y), \text{ with } x = k*i, \; y = k*j,$$
$$i = 1, 2, \ldots, M/k - 1, \; j = 1, 2, \ldots, N/k - 1\}.$$

Similarly we form vectors $\mathbf{f}_{ar}$, $\mathbf{f}_{br}$ and $\mathbf{f}_{bl}$, and the observation vectors for these pixels in a four-block boundary $\mathbf{g}_{al}, \mathbf{g}_{ar}, \mathbf{g}_{br}$ and $\mathbf{g}_{bl}$ (see figure 1.1). Using the vectors above, we also define

$$\mathbf{f}_x^t = \{(\mathbf{f}_{al}(i) \; \mathbf{f}_{ar}(i) \; \mathbf{f}_{br}(i) \; \mathbf{f}_{bl}(i), \; i = 1, \ldots, m)\},$$
$$\mathbf{g}_x^t = \{(\mathbf{g}_{al}(i) \; \mathbf{g}_{ar}(i) \; \mathbf{g}_{br}(i) \; \mathbf{g}_{bl}(i), \; i = 1, \ldots, m)\},$$

with

$$m = (M/k - 1) \times (N/k - 1). \qquad (1.4)$$

We shall then use

$$\mathbf{f}^t = (\mathbf{f}_c^t \ \mathbf{f}_r^t \ \mathbf{f}_x^t) \ \text{ and } \ \mathbf{g}^t = (\mathbf{g}_r^t \ \mathbf{g}_c^t \ \mathbf{g}_x^t),$$

when needed. *Note that originally* $\mathbf{f}$ *represented the complete original image but since the method we are proposing only modifies the pixels at the boundaries we can redefine* $\mathbf{f}$ *as shown above.*

In order to capture the vertical local properties of the image we define a $p \times p$ diagonal matrix, $\mathbf{W}_c$, with p defined in Eq. (1.2), of the form [78]

$$\mathbf{W}_c = \begin{bmatrix} \omega_c(1) & 0 & 0 & \cdots & 0 \\ 0 & \omega_c(2) & 0 & \cdots & \vdots \\ 0 & 0 & \ddots & 0 & \vdots \\ \vdots & \vdots & 0 & \ddots & 0 \\ 0 & \cdots & \cdots & 0 & \omega_c(p) \end{bmatrix}, \tag{1.5}$$

where the $\omega_c(i)$'s, $i = 1, 2, \ldots, p$, determine the relative importance of the intensity differences as shown in figure 1.2 (note that the pixels in a four block boundary are not weighted by this matrix). Analogously, we define $\mathbf{W}_r$ to capture the horizontal local properties of the image, as shown in figure 1.2. The size of this matrix is $q \times q$, where q is defined in Eq. (1.3) (see [78, 39] for details).

For pixels in a four block intersection we take into account the differences $\mathbf{f}_{al} - \mathbf{f}_{ar}$, $\mathbf{f}_{ar} - \mathbf{f}_{br}$, $\mathbf{f}_{br} - \mathbf{f}_{bl}$ and $\mathbf{f}_{bl} - \mathbf{f}_{al}$, corresponding to column, row, column and row differences, respectively (see figures 1.1 and 1.2). The corresponding weight matrices are denoted by $\mathbf{W}_{x1}, \mathbf{W}_{x2}, \mathbf{W}_{x3}$ and $\mathbf{W}_{x4}$, with entries w_c, w_r, w_c and w_r (see figure 1.2). All these matrices have size $m \times m$, where m is defined in Eq. (1.4).

Let us now consider the adaptive image model we will be using. In this chapter we assume that the degree of smoothness in the vertical and horizontal directions is the same, resulting in the use of one smoothing parameter (see [40] for extensions to the use of two parameters).

With the above definitions, the prior knowledge about the smoothness of the block boundaries of the image has the form

$$p(\mathbf{f} \mid \alpha) \propto \alpha^{\frac{p+q+3m}{2}} \exp\left\{-A(\mathbf{f} \mid \alpha)\right\}, \tag{1.6}$$

with

$$\begin{aligned} A(\mathbf{f} \mid \alpha) = \ & \alpha \parallel \mathbf{W}_c(\mathbf{f}_{cl} - \mathbf{f}_{cr}) \parallel^2 + \alpha \parallel \mathbf{W}_r(\mathbf{f}_{ra} - \mathbf{f}_{rb}) \parallel^2 \\ & + \ \frac{1}{2}\alpha \left\{ \parallel \mathbf{W}_{x1}(\mathbf{f}_{al} - \mathbf{f}_{ar}) \parallel^2 + \parallel \mathbf{W}_{x2}(\mathbf{f}_{ar} - \mathbf{f}_{br}) \parallel^2 \right. \\ & + \ \parallel \mathbf{W}_{x3}(\mathbf{f}_{br} - \mathbf{f}_{bl}) \parallel^2 + \parallel \mathbf{W}_{x4}(\mathbf{f}_{bl} - \mathbf{f}_{al}) \parallel^2 \left. \right\}, \tag{1.7} \end{aligned}$$

where α measures the roughness between two block boundaries.

Fidelity to the data at these block boundaries is expressed by

$$p(\mathbf{g} \mid \mathbf{f}, \beta) \propto \beta^{p+q+2m} \exp \left\{ -N(\mathbf{g} \mid \mathbf{f}, \beta) \right\}, \qquad (1.8)$$

where $N(\mathbf{g} \mid \mathbf{f}, \beta)$ is defined by

$$
\begin{aligned}
N(\mathbf{g} \mid \mathbf{f}, \beta) \;=\; & \frac{1}{2}\beta \left\{ \| \mathbf{g}_{cl} - \mathbf{f}_{cl} \|^2 + \| \mathbf{g}_{cr} - \mathbf{f}_{cr} \|^2 + \| \mathbf{g}_{ra} - \mathbf{f}_{ra} \|^2 \right. \\
+ \;& \| \mathbf{g}_{rb} - \mathbf{f}_{rb} \|^2 + \| \mathbf{g}_{al} - \mathbf{f}_{al} \|^2 + \| \mathbf{g}_{ar} - \mathbf{f}_{ar} \|^2 \\
+ \;& \left. \| \mathbf{g}_{br} - \mathbf{f}_{br} \|^2 + \| \mathbf{g}_{bl} - \mathbf{f}_{bl} \|^2 \right\},
\end{aligned}
\qquad (1.9)
$$

and β is defined as $\beta^{-1} = \sigma^2_{noise}$, the noise variance in the image.

1.5 PROPOSED ALGORITHM

If α and β are known, then following the Bayesian paradigm it is customary to select, as the reconstruction of $\mathbf{f}$, the image $\mathbf{f}^{(\alpha,\beta)}$ defined by

$$\mathbf{f}^{(\alpha,\beta)} = \arg\min_{\mathbf{f}} \left\{ M(\mathbf{f}, \mathbf{g} \mid \alpha, \beta) \right\}, \qquad (1.10)$$

where

$$M(\mathbf{f}, \mathbf{g} \mid \alpha, \beta) = A(\mathbf{f} \mid \alpha) + N(\mathbf{g} \mid \mathbf{f}, \beta). \qquad (1.11)$$

Differentiating Eq. (1.10) with respect to $\mathbf{f}$ and setting the derivate equal to zero, results in

$$2\alpha \mathbf{W}_c^t \mathbf{W}_c (\mathbf{f}_{cl} - \mathbf{f}_{cr}) + \beta(\mathbf{f}_{cl} - \mathbf{g}_{cl}) = 0 \qquad (1.12)$$

$$2\alpha \mathbf{W}_c^t \mathbf{W}_c (\mathbf{f}_{cr} - \mathbf{f}_{cl}) + \beta(\mathbf{f}_{cr} - \mathbf{g}_{cr}) = 0 \qquad (1.13)$$

$$2\alpha \mathbf{W}_r^t \mathbf{W}_r (\mathbf{f}_{ra} - \mathbf{f}_{rb}) + \beta(\mathbf{f}_{ra} - \mathbf{g}_{ra}) = 0 \qquad (1.14)$$

$$2\alpha \mathbf{W}_r^t \mathbf{W}_r (\mathbf{f}_{rb} - \mathbf{f}_{ra}) + \beta(\mathbf{f}_{rb} - \mathbf{g}_{rb}) = 0 \qquad (1.15)$$

$$\alpha[\mathbf{W}_{x1}^t \mathbf{W}_{x1}(\mathbf{f}_{al} - \mathbf{f}_{ar}) + \mathbf{W}_{x4}^t \mathbf{W}_{x4}(\mathbf{f}_{al} - \mathbf{f}_{bl})] + \beta(\mathbf{f}_{al} - \mathbf{g}_{al}) = 0 \quad (1.16)$$

$$\alpha[\mathbf{W}_{x1}^t \mathbf{W}_{x1}(\mathbf{f}_{ar} - \mathbf{f}_{al}) + \mathbf{W}_{x2}^t \mathbf{W}_{x2}(\mathbf{f}_{ar} - \mathbf{f}_{br})] + \beta(\mathbf{f}_{ar} - \mathbf{g}_{ar}) = 0 \quad (1.17)$$

$$\alpha[\mathbf{W}_{x3}^t \mathbf{W}_{x3}(\mathbf{f}_{br} - \mathbf{f}_{bl}) + \mathbf{W}_{x2}^t \mathbf{W}_{x2}(\mathbf{f}_{br} - \mathbf{f}_{ar})] + \beta(\mathbf{f}_{br} - \mathbf{g}_{br}) = 0 \quad (1.18)$$

$$\alpha[\mathbf{W}_{x3}^t \mathbf{W}_{x3}(\mathbf{f}_{bl} - \mathbf{f}_{br}) + \mathbf{W}_{x4}^t \mathbf{W}_{x4}(\mathbf{f}_{bl} - \mathbf{f}_{al})] + \beta(\mathbf{f}_{bl} - \mathbf{g}_{bl}) = 0. \quad (1.19)$$

From Eqs. (1.12)–(1.15) we obtain

$$\mathbf{f}_{cl}(i) + \mathbf{f}_{cr}(i) = \mathbf{g}_{cl}(i) + \mathbf{g}_{cr}(i),$$

$$\mathbf{f}_{ra}(i) + \mathbf{f}_{rb}(i) = \mathbf{g}_{ra}(i) + \mathbf{g}_{rb}(i),$$

$$\mathbf{f}_{cl}(i) - \mathbf{f}_{cr}(i) = \gamma_c(i)(\mathbf{g}_{cl}(i) - \mathbf{g}_{cr}(i)),$$

$$\mathbf{f}_{ra}(i) - \mathbf{f}_{rb}(i) = \gamma_r(i)(\mathbf{g}_{ra}(i) - \mathbf{g}_{rb}(i)),$$

with

$$\gamma_c(i) = \frac{\beta}{4\alpha\omega_c^2(i) + \beta},$$

$$\gamma_r(i) = \frac{\beta}{4\alpha\omega_r^2(i) + \beta}.$$

Then we obtain the new values for $\mathbf{f}_{cl}$, $\mathbf{f}_{cr}$, $\mathbf{f}_{ra}$ and $\mathbf{f}_{rb}$ by

$$\mathbf{f}_{cl}(i) = \frac{1}{2}(1 + \gamma_c(i))\,\mathbf{g}_{cl}(i) + \frac{1}{2}(1 - \gamma_c(i))\,\mathbf{g}_{cr}(i) \tag{1.20}$$

$$\mathbf{f}_{cr}(i) = \frac{1}{2}(1 - \gamma_c(i))\,\mathbf{g}_{cl}(i) + \frac{1}{2}(1 + \gamma_c(i))\,\mathbf{g}_{cr}(i) \tag{1.21}$$

$$\mathbf{f}_{ra}(i) = \frac{1}{2}(1 + \gamma_r(i))\,\mathbf{g}_{ra}(i) + \frac{1}{2}(1 - \gamma_r(i))\,\mathbf{g}_{rb}(i) \tag{1.22}$$

$$\mathbf{f}_{rb}(i) = \frac{1}{2}(1 - \gamma_r(i))\,\mathbf{g}_{ra}(i) + \frac{1}{2}(1 + \gamma_r(i))\,\mathbf{g}_{rb}(i). \tag{1.23}$$

$\mathbf{f}_{al}$, $\mathbf{f}_{ar}$, $\mathbf{f}_{br}$ and $\mathbf{f}_{bl}$ are obtained in a similar fashion by solving the equation system described by Eqs. (1.16)–(1.19).

It is obvious that in order to estimate the original image, that is, in order to solve Eq. (1.10) we need to know or estimate the unknown hyperparameters. Yang *et al.* [77, 78] proposed empirical procedures to estimate these parameters within the POCS and CLS approaches to the reconstruction problem. As we have already mentioned, in this work we follow the so called hierarchical Bayesian paradigm, according to which the unknown hyperparameters are estimated by integrating $p(\alpha, \beta, \mathbf{f}, \mathbf{g})$, defined in Eq. (1.1), over $\mathbf{f}$ to obtain $p(\alpha, \beta \mid \mathbf{g})$, which is then maximized over the hyperparameters. If prior knowledge about the unknown hyperparameters is available it can be incorporated into the estimation procedure.

When the hyperparameters α and β are estimated at the decoder but there is no prior information about them, what is needed is a *non-informative* prior (the term *non-informative* is meant to imply that no information about the hyperparameters is contained in the priors). For the problem at hand we can use improper non informative priors of the form

$$p(\omega) \propto const \text{ over } [0, \infty), \tag{1.24}$$

where ω denotes a hyperparameter. With these hyperpriors, the proposed method for estimating α and β amounts to selecting $\hat{\alpha}$ and $\hat{\beta}$, as the maximum likelihood estimates of α and β from $p(\mathbf{g} \mid \alpha, \beta)$.

Let us describe the estimation process in detail. Let us fix α and β and expand $M(\mathbf{f}, \mathbf{g} \mid \alpha, \beta)$ in Eq. (1.11) around $\mathbf{f}^{(\alpha,\beta)}$. We then have,

$$p(\alpha, \beta \mid \mathbf{g}) \propto p(\mathbf{g} \mid \alpha, \beta) \propto \frac{\exp\left\{-M(\mathbf{f}^{(\alpha,\beta)}, \mathbf{g} \mid \alpha, \beta)\right\}}{\alpha^{-\frac{p+q+3m}{2}}\beta^{-(p+q+2m)}}$$

$$\times \quad \int_{\mathbf{f}} \exp\left\{-\frac{1}{2}(\mathbf{f} - \mathbf{f}^{(\alpha,\beta)})^t \mathbf{Q}(\alpha,\beta)(\mathbf{f} - \mathbf{f}^{(\alpha,\beta)})\right\} d\mathbf{f}$$

$$= \quad \frac{\exp\left\{-M(\mathbf{f}^{(\alpha,\beta)}, \mathbf{g} \mid \alpha, \beta)\right\}}{\alpha^{-\frac{p+q+3m}{2}}\beta^{-(p+q+2m)}}(\det \mathbf{Q}(\alpha,\beta))^{-1/2}, \quad (1.25)$$

where $\mathbf{Q}(\alpha,\beta)$ is a $(2p + 2q + 4m) \times (2p + 2q + 4m)$ block diagonal matrix which consists of p matrices of order 2×2 defined by

$$\mathbf{q}_{(i,i)}(\alpha,\beta) = \mathbf{q}_{c(i,i)}(\alpha,\beta) = \alpha \mathbf{A}_c(i) + \beta \mathbf{I}_{2\times 2} \quad (i = 1, \ldots, p),$$

where

$$\mathbf{A}_c(i) = \begin{bmatrix} 2\omega_c^2(i) & -2\omega_c^2(i) \\ -2\omega_c^2(i) & 2\omega_c^2(i) \end{bmatrix},$$

q matrices of order 2×2 defined by

$$\mathbf{q}_{(i,i)}(\alpha,\beta) = \mathbf{q}_{r(i,i)}(\alpha,\beta) = \alpha \mathbf{A}_r(i) + \beta \mathbf{I}_{2\times 2} \quad (i = 1, \ldots, q),$$

where

$$\mathbf{A}_r(i) = \begin{bmatrix} 2\omega_r^2(i) & -2\omega_r^2(i) \\ -2\omega_r^2(i) & 2\omega_r^2(i) \end{bmatrix},$$

and m matrices of order 4×4 defined by

$$\mathbf{q}_{(i,i)}(\alpha,\beta) = \mathbf{q}_{x(i,i)}(\alpha,\beta) = \alpha \mathbf{A}_x(i) + \beta \mathbf{I}_{4\times 4} \quad (i = 1, \ldots, m),$$

where

$$\mathbf{A}_x(i) = \begin{bmatrix} \omega_{x1}^2(i) + \omega_{x4}^2(i) & -\omega_{x1}^2(i) & 0 & -\omega_{x4}^2(i) \\ -\omega_{x1}^2(i) & \omega_{x1}^2(i) + \omega_{x3}^2(i) & -\omega_{x3}^2(i) & 0 \\ 0 & -\omega_{x2}^2(i) & \omega_{x2}^2(i) + \omega_{x4}^2(i) & -\omega_{x4}^2(i) \\ -\omega_{x1}^2(i) & 0 & -\omega_{x3}^2(i) & \omega_{x2}^2(i) + \omega_{x3}^2(i) \end{bmatrix}.$$

Then, differentiating $-\log p(\alpha, \beta \mid \mathbf{g})$ with respect to α and β we have

$$\frac{p+q+3m}{\alpha} = 2 \parallel \mathbf{W}_c(\mathbf{f}_{cl}^{(\alpha,\beta)} - \mathbf{f}_{cr}^{(\alpha,\beta)}) \parallel^2 + 2 \parallel \mathbf{W}_r(\mathbf{f}_{ra}^{(\alpha,\beta)} - \mathbf{f}_{rb}^{(\alpha,\beta)}) \parallel^2$$

$$+ \parallel \mathbf{W}_{x1}(\mathbf{f}_{al}^{(\alpha,\beta)} - \mathbf{f}_{ar}^{(\alpha,\beta)}) \parallel^2 + \parallel \mathbf{W}_{x2}(\mathbf{f}_{ar}^{(\alpha,\beta)} - \mathbf{f}_{br}^{(\alpha,\beta)}) \parallel^2$$

$$+ \parallel \mathbf{W}_{x3}(\mathbf{f}_{br}^{(\alpha,\beta)} - \mathbf{f}_{bl}^{(\alpha,\beta)}) \parallel^2 + \parallel \mathbf{W}_{x4}(\mathbf{f}_{bl}^{(\alpha,\beta)} - \mathbf{f}_{al}^{(\alpha,\beta)}) \parallel^2$$

$$+ \sum_{i=1}^{p} \text{trace} \left[[\mathbf{q}_{c(i,i)}(\alpha,\beta)]^{-1}\mathbf{A}_c(i)\right]$$

$$+ \sum_{i=1}^{q} \text{trace} \left[[\mathbf{q}_{r(i,i)}(\alpha,\beta)]^{-1}\mathbf{A}_r(i)\right]$$

$$+ \sum_{i=1}^{m} \text{trace}\left[\left[\mathbf{q}_{x(i,i)}(\alpha,\beta)\right]^{-1}\mathbf{A}_x(i)\right], \tag{1.26}$$

$$\frac{2(p+q+2m)}{\beta} = \| \mathbf{f}_{cl}^{(\alpha,\beta)} - \mathbf{g}_{cl} \|^2 + \| \mathbf{f}_{cr}^{(\alpha,\beta)} - \mathbf{g}_{cr} \|^2 + \| \mathbf{f}_{ra}^{(\alpha,\beta)} - \mathbf{g}_{ra} \|^2$$

$$+ \| \mathbf{f}_{rb}^{(\alpha,\beta)} - \mathbf{g}_{rb} \|^2 + \| \mathbf{f}_{al}^{(\alpha,\beta)} - \mathbf{g}_{al} \|^2 + \| \mathbf{f}_{ar}^{(\alpha,\beta)} - \mathbf{g}_{ar} \|^2$$

$$+ \| \mathbf{f}_{bl}^{(\alpha,\beta)} - \mathbf{g}_{bl} \|^2 + \| \mathbf{f}_{br}^{(\alpha,\beta)} - \mathbf{g}_{br} \|^2$$

$$+ \sum_{i=1}^{p+q+m} \text{trace}\left[\left[\mathbf{q}_{(i,i)}(\alpha,\beta)\right]^{-1}\right]. \tag{1.27}$$

Therefore, by solving Eqs. (1.26) and (1.27) an estimate of the unknown hyperparameters is obtained. We note that in the process of estimating the hyperparameters we also need to calculate the corresponding $\mathbf{f}^{(\alpha,\beta)}$.

In summary, the following iterative algorithm can be used to recover $\mathbf{f}$, an image with reduced blocking artifacts and to estimate the needed hyperparameters.

> **ALGORITHM 1** {
>
> Choose α^0 and β^0.
>
> Compute $\mathbf{f}^{(\alpha^0,\beta^0)}$ from Eqs. (1.20)–(1.23) and by solving the system of equations described by Eqs. (1.16)–(1.19).
>
> Set $k = 1$.
>
> Repeat {
>
> Estimate α^k and β^k by substituting α^{k-1} and β^{k-1} in the right hand side of Eqs. (1.26) and (1.27), respectively.
>
> Compute $\mathbf{f}^{(\alpha^k,\beta^k)}$ from Eqs. (1.20)–(1.23) and by solving the system of equations described by Eqs. (1.16)–(1.19).
>
> } until $\| \mathbf{f}^{(\alpha^k,\beta^k)} - \mathbf{f}^{(\alpha^{k-1},\beta^{k-1})} \|$ is less than a prescribed bound.
>
> }

The convergence of this method is established by noting that it corresponds to the EM algorithm with complete data the observations $\mathbf{g}$ and the unknown reconstruction $\mathbf{f}$, that is $\mathbf{z}^t = (\mathbf{f}^t, \mathbf{g}^t)$ and

$$\mathbf{g} = (\mathbf{I}\ 0)\, \mathbf{z}$$

(see [45] for details).

It is clear that this process for estimating the image and the hyperparameters can also be performed at the coder. In this case Eq. (1.10) becomes

$$\begin{aligned}
\mathbf{f}_{(\alpha,\beta)}^c &= \arg\min_{\mathbf{p}} \left\{ M(\mathbf{p}, \mathbf{f} \mid \alpha, \beta) \right\} \\
&= \arg\min_{\mathbf{p}} \left\{ A(\mathbf{p} \mid \alpha) + N(\mathbf{f} \mid \mathbf{p}, \beta) \right\},
\end{aligned} \tag{1.28}$$

and the iterative procedure described in Algorithm 1 can also be applied here. We denote by α^c and β^c the hyperparameters obtained at the coder by the previously described iterative procedure using the original image.

1.6 COMBINING INFORMATION FROM THE CODER

Let us now assume that (quantized) versions of the hyperparameters α and β are received by the decoder, denoted respectively by m^c and n^c, and that we want to use these values as prior information in guiding the estimation of the hyperparameters at the decoder. To do so formally within the Bayesian paradigm, we use the following gamma hyperpriors for each hyperparameter

$$p(\alpha) \quad \propto \quad \alpha^{l(m^c)-1} \exp[-l(m^c)\alpha/m^c], \tag{1.29}$$

$$p(\beta) \quad \propto \quad \beta^{l(n^c)-1} \exp[-l(n^c)\beta/n^c]. \tag{1.30}$$

Note that the gamma distribution is defined by

$$p(\omega) \propto \omega^{l-1} \exp[-al\omega],$$

where $l > 1$, ω denotes a hyperparameter and a and l are explained below. Since the gamma distribution has the following properties

$$E[w] = a^{-1} \quad and \quad Var[w] = [a^2 l]^{-1}, \tag{1.31}$$

the mean of w, which represents the inverse of the prior or noise variance, is equal to $1/a$, and its variance decreases when l increases. So, l can then be understood as a measure of the certainty on the knowledge about the prior or noise variances.

We follow again the hierarchical Bayesian approach to the reconstruction problem. We now use, in Eq. (1.1), the gamma distributions defined in Eqs. (1.29) and (1.30), and the distributions in Eqs. (1.6) and (1.8) as the image and noise models, respectively. We obtain the following iterative procedure for recovering $\mathbf{f}$, an image with reduced blocking artifacts, and for estimating the unknown hyperparameters with the information provided by the coder (see [40, 37] for details)

> **ALGORITHM 2** {
>
>   Set $\mathbf{f}^0 = \mathbf{g}$.
>   Choose α^0 and β^0.
>   Set $k = 1$.
>   Repeat {
>   • Estimate α^d and β^d by substituting α^{k-1}, β^{k-1} and their corresponding MAP solution, $\mathbf{f}^{(\alpha^{k-1},\beta^{k-1})}$, in the right hand side of Eqs. (1.26) and (1.27), respectively.
>   • Set

$$(\alpha^k)^{-1} = \mu(m^c)^{-1} + (1 - \mu)(\alpha^d)^{-1},$$

and

$$(\beta^k)^{-1} = \nu(n^c)^{-1} + (1 - \nu)(\beta^d)^{-1}.$$

where

$$\mu = \frac{l(m^c) - 1}{l(m^c) - 1 + (p + q + 3m)/2} \qquad (1.32)$$

and

$$\nu = \frac{l(n^c) - 1}{l(n^c) - 1 + p + q + 2m} \qquad (1.33)$$

- Compute $\mathbf{f}^{(\alpha^k, \beta^k)}$, that is, $\mathbf{f}_{cl}^k$, $\mathbf{f}_{cr}^k$, $\mathbf{f}_{ra}^k$, $\mathbf{f}_{rb}^k$, $\mathbf{f}_{al}^k$, $\mathbf{f}_{ar}^k$, $\mathbf{f}_{br}^k$ and $\mathbf{f}_{bl}^k$ from Eqs. (1.16)–(1.23).
  } until $\| \mathbf{f}^{(\alpha^k, \beta^k)} - \mathbf{f}^{(\alpha^{k-1}, \beta^{k-1})} \|$ is less than a prescribed bound.
}

Note that μ can be considered as a normalized confidence parameter since it takes values in the interval $[0, 1)$. It is clear that when $\mu = 0$ no confidence is put in the estimate of the prior variance at the coder, while $\mu = 1$ fully enforces the prior knowledge of the prior variance. We also note that $\mu = 1$ corresponds to $l(m^c) = \infty$, that is, the corresponding gamma distribution has zero variance (see Eq. (1.31)). The same ideas can be used to explain the meaning of ν.

It is clear that the way the parameters are weighted depends on how they improve the peak signal to noise ratio and on the way they are transmitted. For instance, if the parameters are transmitted without loss and they result in a higher peak signal to noise ratio, then they should be used instead of the ones obtained at the decoder. However, if they are quantized, better results could be obtained by using a convex combination with the ones estimated at the decoder. This will be examined in section 1.8.

1.7 EXTENSION TO COLOR IMAGES

Most of the developed algorithms to reduce the blocking artifact deal only with grey scale images, and if applied to color images, they only process the Y (luminance) band in the YCbCr format, that is, they do not modify the Cb and Cr bands.

In the JPEG specifications for color images [51] each band is coded independently, in other words, the same coding is performed on the chrominance and two luminance bands, with the use of different quantization tables. A subsampling technique is also applied to the chrominance bands, usually noted as 4:2:2 or 4:1:1. The 4:2:2 format specifies that for every four samples of Y information there are two samples of Cb and Cr, while the 4:1:1 format, specifies that for every four samples of Y, there is one sample of Cb and Cr.

Let $\mathbf{f}$ now be a three band color image, whose components will be denoted by $\mathbf{f}^I$, $I \in \{Y, Cb, Cr\}$, and let $\mathbf{g}$ be the corresponding compressed image. Let $U \times V$ be the original size of each band and assume that $k \times k$ blocks are used to codify each band, where U and V are multiples of k.

For a 4:1:1 subsampling, the one used in this chapter, we note that for the luminance band, Y, a $M(Y) \times N(Y)$ image is processed, with $M(Y) = U$ and $N(Y) = V$, and for the chrominance bands, Cb and Cr, the images are $M(I) \times N(I)$, where $M(I) = U/2$ and $N(I) = V/2$, with $I \in \{Cb, Cr\}$. However, in both cases the blocks used are $k \times k$.

We now apply independently to each band all the proposed reconstruction algorithms and then define the reconstruction for the three band problem as the three band image obtained by combining the $\{Y, Cb, Cr\}$ reconstructions (see [38] for details). In the next section we compare the results obtained with this approach and the standard approach of processing the Y band only.

Finally, we note that although we are processing each band independently, chrominance information can be used to reconstruct the chrominance band, as was done in luminance [70].

1.8 TEST EXAMPLES

In this section, experiments are presented in order to test the describe recovery algorithms. The peak signal-to-noise ratio ($PSNR$) is used to evaluate the quality of the reconstructed image. It is defined by

$$PSNR = 10 \log_{10} \frac{255^2}{\|\mathbf{f} - \hat{\mathbf{f}}\|^2},$$

where $\mathbf{f}$ denotes the original image and $\hat{\mathbf{f}}$ its reconstruction (when no post-processing is performed $\hat{\mathbf{f}} = \mathbf{g}$). The 512×512 "Lena" image was used in the reported experiments. The image was first compressed using a JPEG based coder-decoder at a bit-rate of 0.22 bpp. For presentation purposes the center 256×256 section of this image is shown in figure 1.3 having a $PSNR$ of 29.58 dB. The reconstructed image using the proposed algorithm with the hyperparameters estimated only at the decoder is shown in figure 1.3 ($PSNR = 30.16$ dB). The estimated values of α_d^{-1} and β_d^{-1} were equal to 238.8 and 32.9, respectively.

The values of the hyperparameters estimated from the original image were $\alpha_c^{-1} = 29.3$ and $\beta_c^{-1} = 22.6$. For these values of the hyperparameters, the $PSNR$ was equal to 30.37 dB.

We then proceeded to combine the hyperparameters estimated at the coder and the decoder, using Algorithm 2. The combination is obtained with the use of the parameters μ and ν in Eqs. (1.32) and (1.33). These parameters range from 1 to 0, from more to less confidence on the estimation at the coder. We used $\mu = \nu$ in our experiments, with a step of 0.1. Figure 1.4 shows how the values of α change as a function of μ and ν. The corresponding evolution of β is shown in figure 1.5. The way the $PSNR$ changes as a function of μ and ν is shown in figure 1.6. Note that the values of α and β shown in figures 1.4 and 1.5 are not linear combinations of the ones obtained at the coder (with the original image) and the ones obtained at the decoder (flat hyperprior and compressed image). These experiments seem to suggest that, for this compression level, the best results are obtained when we estimate the hyperparameters from the original image.

We then performed the same experiments with quantized versions of the hyperparameters estimated at the coder. The quantization value of 10 as used, which means that the values of the hyperparameters were $\alpha_c^{-1} = 30$ and $\beta_c^{-1} = 20$. The same experiments as above were performed; the evolution of α, β and $PSNR$ is shown in figures 1.4, 1.5 and 1.6, respectively. As demonstrated in figure 1.6, these experiments seem to imply that when quantized versions of the original hyperparameters are used it is still better to estimate the hyperparameters from the original image.

The same experiments were performed at a bit rate of 0.36 bpp, with a $PSNR$ of 30.96 dB. The reconstructed image using the proposed algorithm with the hyperparameters estimated only at the decoder is shown in figure 1.7 ($PSNR = 44.35$ dB). The values of α_d^{-1} and β_d^{-1} were equal to 77.47 and 19.93, respectively.

Using the corresponding values of the hyperparameters estimated from the original image, the corresponding $PSNR$ was equal to 40.95 dB, which seems to suggest that for this compression level better results are obtained when the parameters are estimated at the decoder.

We then proceeded again to combine the hyperparameters at the coder with the ones at the decoder using algorithm 2. This combination is obtained with the use of the parameters μ and ν ranging from 1 to 0. We used again $\mu = \nu$ with a step of 0.1. Figures 1.8, 1.9 and 1.10 show respectively α, β and $PSNR$ as a function of $\mu = \nu$. These experiments seem to suggest that, in this case, the best results are obtained when we estimate the hyperparameters from the compressed image. We note that an exhaustive study of the methods which estimate the parameters at the coder and the decoder, as well as the combination of these parameters, including the use of quantized versions of them, is carried out in [37].

We then performed the same experiments with quantized versions of the hyperparameters estimated at the coder. The quantization value of 10 was used which means that the values of the hyperparameters were $\alpha_c^{-1} = 30$ and $\beta_c^{-1} = 20$. The evolution of α, β and $PSNR$ are shown in figures 1.8, 1.9 and 1.10, respectively. These experiments seem to imply that when quantized versions of the original hyperparameters are used it is better to estimate the hyperparameters from the compressed image.

Finally, we performed a simple experiment with a color image. The color "Lena" image was used and the same central region is used for display. In figure 1.11 we show the Cb and Cr bands of the "Lena" image and the reconstruction of the bands when the method described in section 1.7 is applied to the color image. The original, compressed and reconstructed color images can be retrieved from *http://decsai.ugr.es/~jmd/book_chapter*. All parameters were estimated at the decoder. It is clear from figure 1.11 and by observing the actual color images that the quality of the reconstructed images improves when the chrominance bands are reconstructed as well.

1.9 CONCLUSIONS

In this chapter a survey of the literature to remove blocking artifacts has been carried out. A new spatially-adaptive image recovery algorithm based on the Bayesian hierarchical approach has been described in detail to decode BDCT based compressed image. Using this approach we have shown how to estimate the unknown hyperparameters using well grounded estimation procedures and how to incorporate from vague to precise knowledge about the unknown parameters into the recovery process. The performed tests show good improvement in term of the $PSNR$ metric and the visual quality of the images.

References

[1] G. Archer and D. M. Titterington. On some bayesian/regularization methods for image restoration. *IEEE Trans. on Image Processing*, 4:989–995, 1995.

[2] A. Baskurt, R. Prost, and R. Goutte. Iterative constrained restoration of dct-compressed images. *Signal Processing*, 17:201–211, 1989.

[3] J. O. Berger. *Statistical Decision Theory and Bayesian Analysis*. New York, Springer Verlag, 1985.

[4] J. C. Brailean, T. Özcelik, and A. K. Katsaggelos. A video coding algorithm based on recovery techniques using mean field annealing. In *Proceedings of the Visual Communication and Image Processing 95, SPIE Proc.*, pages 284–294, 1995.

[5] W. L. Buntine. *A Theory of Learning Classification Rules*. PhD thesis, University of Technology, Sydney, Australia, 1991.

[6] W. L. Buntine. Theory refinement on Bayesian networks. In *Proc. of the Seventh Conference on Uncertainty in Artificial Intelligence*, pages 52–60, 1991.

[7] W. L. Buntine and A. Weigund. Bayesian back-propagation. *Complex Systems*, 5:603–643, 1991.

[8] Y.-H. Chan and W.-C. Siu. An approach to subband DCT image coding. *Journal of Visual Communication and Image Representation*, 5(1):95–106, March 1994.

[9] S. S. O. Choy, Y.-H. Chan, and W.-C. Siu. Reduction of block-transform image coding artifacts by using local statistics of transform coefficients. *IEEE Signal Processing Letters*, 4(1):5–7, January 1997.

[10] G. F. Cooper and E. Herkovsits. A bayesian method for the induction of probabilistic networks from data. *Machine Learning*, 9:309–347, 1992.

[11] M. Crouse and K Ramchandran. Nonlinear constrained least squares estimation to reduce artifacts in block transform-coded images. In *Proceedings of the International Conference on Image Processing ICIP 95*, volume 1, pages 462–465, 1995.

[12] C. Derviaux, F. X. Coudoux, M. G. Gazalet, and P. Corlay. Blocking artifact reduction of DCT coded image sequences using a visually adaptive postprocessing. In *Proceedings of the International Conference on Image Processing ICIP 96*, volume 2, pages 5–8, 1996.

[13] Y.-H. Fok, O. C. Au, and C. Chang. Bitrate and blocking artifact reduction by iterative pre-distorsion. In *Proceedings of the International Conference on Image Processing ICIP 96*, volume 2, pages 13–16, 1996.

[14] S. F. Gull. Developments in maximum entropy data analysis. In J. Skilling, editor, *Maximum Entropy and Bayesian Methods*, pages 53–71. Kluwer, 1989.

[15] Pong-Sik Ho and Min-Hwan Kim. Perceptually-adaptive image compression based on block dct. *Journal of KISS[A] [Computer Systems and Theory]*, 22(10):1405–1415, 1995.

[16] S. W. Hong, Y. H. Chan, and W. C. Siu. A practical real time postprocessing technique for block effect elimination. In *Proceedings of the International Conference on Image Processing ICIP 96*, volume 2, pages 21–24, 1996.

[17] T.-C. Hsung, D. P.-K. Lun, and W.-C. Siu. A deblocking techniqhe for JPEG decoded image using wavelets transform modulus maxima representation. In *Proceedings of the International Conference on Image Processing ICIP 96*, volume 2, pages 561–564, 1996.

[18] (ISO/IEC). *Digital Compression and Coding of Continuous-tone Still Images, Part 1, Requirements and Guidelines*. ISO/IEC JTC 1 / SC 29 International Standard 10918-1, 1994.

[19] Y. Itoh. Detail preserving noise filtering for compressed image. *IEICE Trans. on Communications*, E79-B(10):1459–1466, October 1996.

[20] Y. Itoh. Detail-preserving noise filtering using binary index. In *Proceedings of the Image and Video Processing IV Conf., SPIE Proc.*, volume 2666, pages 119–130, 1996.

[21] B. Jeon, J. Jeong, and J. M. Jo. Blocking artifacts reduction in image coding based on minimum block boundary discontinuity. In *Proceedings of the Visual Communications and Image Processing 95*, pages 198–209, 1995.

[22] J. Jeong and B. Jeon. Use of a class of two-dimensional functions for blocking artifacts reduction in image coding. In *Proceedings of the International Conference on Image Processing ICIP 95*, volume 1, pages 478–481, 1995.

[23] H. Joung, U. Chong, and S. P. Kim. Block artifact reduction by optimization utilizing subband decomposition. In *Proceedings of the 7th KSEA Northeast Regional Conference*, 1996.

[24] H. C. Kim and H. W. Park. Signal adaptive postprocessing for blocking effects reduction in JPEG image. In *Proceedings of the International Conference on Image Processing ICIP 96*, volume 2, pages 41–44, 1996.

[25] C. Kuo and R. Hsieh. Adaptive postprocessor for block encoded images. *IEEE Trans. on Circuits and Systems for Video Technology*, 5(4):298–304, August 1995.

[26] K. Y. Kwak and R.A Haddad. Projection-based eigenvector decomposition for reduction of blocking artifacts of DCT coded image. In *Proceedings of the International Conference on Image Processing ICIP 95*, volume 2, pages 527–530, 1995.

[27] Y.-K. Lai, J. Li, and C.-C. J. Kuo. Image enhancement for low bit-rate jpeg and mpeg coding via postprocessing. In *Proceedings of the Visual Communications and Image Processing '96, SPIE Proc.*, volume 2727, pages 1484–1494, 1996.

[28] Y.-K. Lai, J. Li, and C.-C. J. Kuo. Removal of blocking artifacts of dct transform by classified space-frequency filtering. In *Conference Record of The Twenty-Ninth Asilomar Conference on Signals, Systems and Computers*, volume 2, pages 1457–1461, 1996.

[29] I. Linares, R. Mersereau, and M. Smith. JPEG estimated spectrum adaptive post-filtering using image adaptive Q-tables and canny edge detectors. In *Proceedings of the 1996 IEEE International Symposium on Circuits and Systems*, volume 2, pages 722–725, 1996.

[30] T.-S. Liu and N. Jayant. Adaptive posprocessing algorithms for low bit rate video signals. *IEEE Trans. on Image Processing*, 4(7):1032–1035, July 1995.

[31] J. Luo, C. W. Chen, K. J. Parker, and T. S. Huang. A new method for block effect removal in low bit-rate image compression. In *Proceedings of the International Conference on Acoustics, Speech, and Signal Processing ICASSP 94*, volume 5, pages 341–344, 1994.

[32] J. Luo, C. W. Chen, K. J. Parker, and T. S. Huang. Artifact reduction in low bit rate DCT-based image compression. *IEEE Trans. on Image Processing*, 5(9):1363–1368, September 1996.

[33] W. E. Lynch, A. R. Reibman, and B. Liu. Post processing transform coded images using edges. In *Proceedings of the 1995 International Conference on Acoustics, Speech, and Signal Processing*, volume 4, 1995.

[34] D. J. C. MacKay. Bayesian interpolation. *Neural Computation*, 4:415–447, 1992.

[35] D. J. C. MacKay. A practical Bayesian framework for backprop networks. *Neural Computation*, 4:448–472, 1992.

[36] D. J. C. MacKay. Hyperparameters: Optimize, or integrate out? Submitted to *Neural Computation*, 1995.

[37] J. Mateos. *Reconstrucción Automática de imágenes comprimidas mediante transformada coseno discreta usando métodos bayesianos*. PhD thesis, E.T.S. Ing. Informática. University of Granada, July 1998.

[38] J. Mateos, C. Ilia, B. Jiménez, R. Molina, and A. K. Katsaggelos. Reduction of blocking artifacts in block transformed compressed color images. In *Proceedings of the International Conference on Image Processing ICIP 98*, 1998. To be held.

[39] J. Mateos, A. K. Katsaggelos, and R. Molina. Parameter estimation in regularized reconstruction of BDCT compressed images for reducing blocking artifacts. In *Proceedings of the Conference on Digital Compression Technologies & Video Communications, SPIE Proc.*, volume 2952, pages 70–81, 1996.

[40] J. Mateos, A. K. Katsaggelos, and R. Molina. A bayesian approach to estimate and transmit regularization parameters for reducing blocking artifacts. Submitted to IEEE Trans on Image Processing, 1998.

[41] J. Mateos, R. Molina, and A. K. Katsaggelos. Estimating and transmitting regularization parameters for reducing blocking artifacts. In *Proceedings of the 13th International Conference on Digital Signal Processing*, pages 209–212, 1997.

[42] J. D. McDonnell, R. Shorten, and A. D. Fagan. An edge classification based approach to the post-processing of transform coded images. In *Proceedings of the International Conference on Acoustics, Speech, and Signal Processing ICASSP 94*, volume 5, pages 329–332, 1994.

[43] S. Minami and A. Zakhor. An optimization approach for removing blocking effects in transform coding. *IEEE Trans. on Circuits and Systems for Video Technology*, 5(2):74–81, April 1995.

[44] R. Molina. On the hierarchical Bayesian approach to image restoration. application to astronomical images. *IEEE Trans. on Pattern Analysis and Machine Intelligence*, 16(11):1222–1228, 1994.

[45] R. Molina, A. K. Katsaggelos, and J. Mateos. Bayesian and regularization methods for hyperparameter estimation in image restoration. Submitted to IEEE Trans on Image Processing, 1997.

[46] T. P. O'Rourke and R. L. Stevenson. Improved image decompression for reduced transform coding artifacts. In *Proceedings of the Image and Video Processing II Conf., SPIE Proc.*, volume 2182, pages 90–101, 1994.

[47] T. P. O'Rourke and R. L. Stevenson. Improved image decompression for reduced transform coding artifacts. *IEEE Trans. on Circuits and Systems for Video Technology*, 5(6):490–499, December 1995.

[48] T. Özcelik, J. C. Brailean, and A. K. Katsaggelos. Image and video compression algorithms based on recovery techniques using mean field annealing. *Proceedings of the IEEE*, 83(2):304–316, February 1995.

[49] H. Paek and S.-U. Lee. A projection-based post-processing technique to reducte blocking artifact using *a prior* information on DCT coefficients of adjacent blocks. In *Proceedings of the International Conference on Image Processing ICIP 96*, volume 2, pages 53–56, 1996.

[50] Hoon Paek, Jong-Wook Park, and Sang-Uk Lee. Non-iterative post-processing technique for transform coded image sequence. In *Proceedings of the International Conference on Image Processing ICIP 95*, volume 3, pages 208–211, 1995.

[51] W. B. Pennebaker and J. L. Mitchell. *JPEG Still Image Compression Standard.* Van Nostrand Reinhold, 1992.

[52] H. A. Peterson, A.J. Ahumada, and A.B Watson. The visibility of DCT quantization noise. In *Society for Information Display. Diggest of technical papers*, pages 942–945, 1993.

[53] H. A. Peterson, A.J. Ahumada, and A.B Watson. Visibility of dct quantization noise: spatial frequency summation. In *Proceedings of the 1994 SID International Symposium Digest of Technical Papers*, pages 704–7, 1994.

[54] R. Prost, Y. Ding, and A. Baskurt. JPEG dequantization array for regularized decompression. *IEEE Trans. on Image Processing*, 6(6):883–888, June 1997.

[55] G. Ramamurthi and A. Gersho. Nonlinear space-variant postprocessing of block coded images. *IEEE Trans. on Acoustic, Speech and Signal Processing*, 34(5):1258–1269, October 1986.

[56] H. C. Reeves and J. S. Lim. Reduction of blocking effects in image coding. *Optical Engineering*, 23(1):34–37, January 1984.

[57] S. J. Reeves and S. L. Eddins. Comments on "iterative procedures for reduction of blocking effects in transform image coding". *IEEE Trans. on Cirucits and Systems for Video Technology*, 3(6):439–440, December 1993.

[58] D. G. Sampson, D. V. Papadimitriou, and G. Chamzas. Post-processing of block-coded images at low bitrates. In *Proceedings of the International Conference on Image Processing ICIP 96*, volume 2, pages 1–4, 1996.

[59] K. Sauer. Enhancement of low bit-rate coded images using edge detection and estimation. *CVGIP: Graphical Models and Image Processing*, 53(1):52–62, January 1991.

[60] D.A. Silverstein and S.A. Klein. Restoration of compressed images. In *Proceedings of the Image and Video Compression Conf., SPIE Proc.*, volume 2186, pages 56–64, 1994.

[61] D. J. Spiegelhalter and S.L. Lauritzen. Sequential updating of conditional probabilities on directed graphical structures. *Networks*, 20:579–605, 1990.

[62] R. L. Stevenson. Reduction of coding artifacts in transform image coding. In *Proceedings of the International Conference on Acoustics, Speech, and Signal Processing ICASSP 93*, volume 5, pages 401–404, 1993.

[63] R. L. Stevenson. Reduction of coding artifacts in low-bit-rate video coding. In *Proceedings of the 38th Midwest Symposium on Circuits and Systems.*, pages 854–857, August 1995.

[64] C. E. M. Strauss, D. H. Wolpert, and D. R. Wolf. Alpha, evidence and the entropic prior. In A. Mohammed-Djafari, editor, *Maximum Entropy and Bayesian Methods, Paris*, pages 53–71. Kluwer, 1992.

[65] A. Sultan and H.A. Latchman. Adaptive quantization scheme for mpeg video coders based on hvs (human visual system). In *Proceedings of the Digital Video Compression: Algorithms and Technologies 1996, SPIE Proc.*, volume 2668, pages 181–188, 1996.

[66] S. Suthaharan and H.R. Wu. Adaptive-neighbourhood image filtering for mpeg-1 coded images. In *Proceedings of the Fourth International Symposium on Signal Processing and its Applications. ISSPA 96*, volume 1, pages 166–167, 1996.

[67] Soon Hie Tan, K. K. Pang, and K. N. Ngan. Classified perceptual coding with adaptive quantization. *IEEE Trans. on Circuits and Systems for Video Technology*, 6(4):375–388, August 1996.

[68] M. Temerinac and B. Edler. Overlapping block transform: Window design, fast algorithm, an image coding experiment. *IEEE Trans. on Communications*, 43(9):2417–2425, 1995.

[69] K.-H. Tzou. Post-filtering of transform-coded images. In *Applications of the digital image processing XI, SPIE Proc.*, volume 974, pages 121–126, 1988.

[70] J. L. Webb. Post-processing to reduce blocking artifacts for low bit-rate video coding using chrominance information. In *Proceedings of the International Conference on Image Processing ICIP 96*, volume 2, pages 9–12, 1996.

[71] D. H. Wolpert. On the use of evidence in neural networks. In C. L. Giles, S. J. Hanson, and J. D. Cowan, editors, *Advances in Neural Information Processing Systems 5, San Mateo, California*, pages 539–546. Morgan Kaufmann, 1993.

[72] Z. Xiong, M. T. Orchard, and Y.-Q. Zhang. A deblocking algorithm for JPEG compressed images using overcomplete wavelet representation. *IEEE Trans. on Circuits and Systems for Video Technology*, 7(4):433–437, April 1997.

[73] Li Yan. Adaptive spatial-temporal postprocessing for low bit-rate coded image sequence. In *Proceedings of the image and video processing II conf., SPIE Proc.*, volume 2182, pages 102–109, 1994.

[74] Li Yan. A nonlinear algorithm for enhancing low bit-rate coded motion video sequence. In *Proceedings of the International Conference on Image Processing ICIP 94*, volume 2, pages 923–927, 1994.

[75] Y. Yang and N. P. Galatsanos. Edge-preserving reconstruction of compressed images using projections and a divide-and-conquer strategy. In *Proceedings of the International Conference on Image Processing ICIP 94*, volume 2, pages 535–539, November 1994.

[76] Y. Yang and N. P. Galatsanos. Compression artifact removal using projections onto convex sets and line process modeling. *IEEE Trans. on Image Processing*, 6(10):1345–1357, October 1997.

[77] Y Yang, N. P. Galatsanos, and A. K. Katsaggelos. Regularized reconstruction to reduce blocking artifacts of block discrete cosine transform compressed images. *IEEE Trans. on Circuits and Systems for Video Technology*, 3(6):421–432, December 1993.

[78] Y. Yang, N. P. Galatsanos, and A. K. Katsaggelos. Projection-based spatially-adaptive reconstruction of block-transform compressed images. *IEEE Trans. on Image Processing.*, 4(7):896–908, July 1995.

[79] A. Zakhor. Iterative procedures fror reduction of blocking effects in transform image coding. *IEEE Trans. on Circuits and Systems for Video Technology*, 2(1):91–95, March 1992.

Figure 1.3 256 × 256 center section of "Lena": (top) JPEG based compression at 0.22 bpp, $PSNR = 29.58$ dB; (bottom) reconstruction, hyperparameters estimated at the decoder, $PSNR = 30.16$ dB.

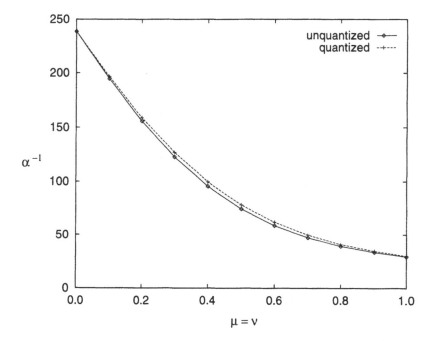

Figure 1.4   Evolution of α^{-1} for the 0.22 bpp image using the parameters estimated at the coder (unquantized and quantized versions of them).

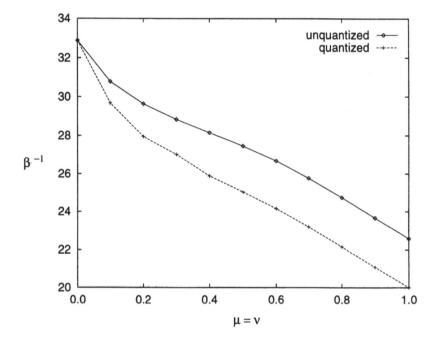

Figure 1.5   Evolution of β^{-1} for the 0.22 bpp image using the parameters estimated at the coder (unquantized and quantized versions of them).

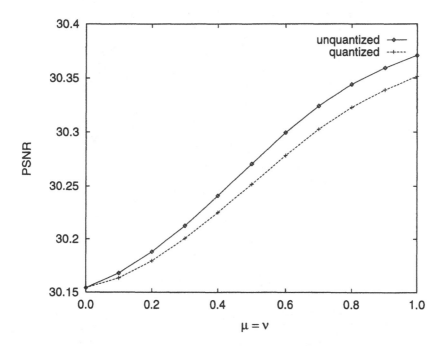

Figure 1.6   Evolution of the $PSNR$ for the 0.22 bpp image using the parameters estimated at the coder (unquantized and quantized versions of them).

Figure 1.7 256×256 center section of "Lena": (top) JPEG based compression at 0.36 bpp, $PSNR = 30.96$ dB; (bottom) reconstruction, hyperparameters estimated at the decoder, $PSNR = 44.35$ dB.

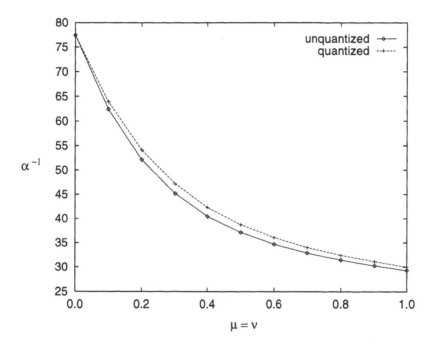

Figure 1.8   Evolution of α^{-1} for the 0.36 bpp image using the parameters estimated at the coder (unquantized and quantized versions of them).

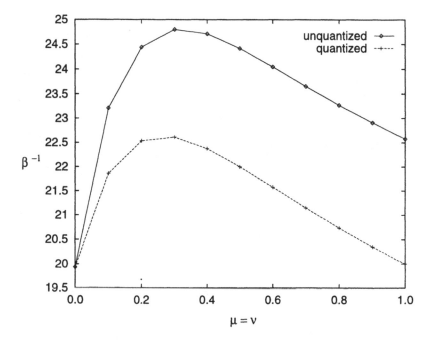

Figure 1.9   Evolution of β^{-1} for the 0.36 bpp image using the parameters estimated at the coder (unquantized and quantized versions of them).

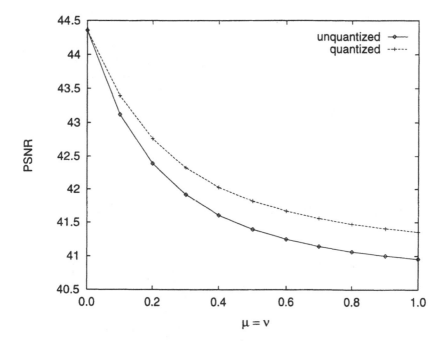

Figure 1.10   Evolution of the $PSNR$ for the 0.36 bpp image using the parameters estimated at the coder (unquantized and quantized versions of them).

34

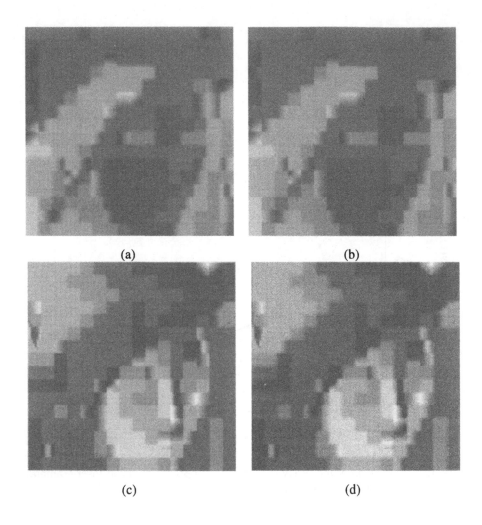

(a)　　　　　　　　　　　　(b)

(c)　　　　　　　　　　　　(d)

Figure 1.11 256×256 center section of color "Lena". All images are scaled to the range [0,255]: (a) JPEG based compression of the Cb band at 0.30 bpp; (b) Cb band reconstruction; (c) JPEG based compression of the Cr band at 0.30 bpp; (d) Cr band reconstruction.

2 A STOCHASTIC TECHNIQUE FOR THE REMOVAL OF ARTIFACTS IN COMPRESSED IMAGES AND VIDEO

Ramon Llados-Bernaus, Mark A. Robertson, and
Robert L. Stevenson

Department of Electrical Engineering
University of Notre Dame
275 Fitzpatrick Hall
Notre Dame, IN 46556
{rlladosb, mrobert2, rls}@nd.edu

Abstract:

The perceived quality of images and video sequences reconstructed from low bit rate compressed bit streams is severely degraded by the appearance of coding artifacts. This chapter introduces a technique for the post-processing of compressed images based on a stochastic model for the image data. Quantization partitions the transform coefficient space and maps all points in a partition cell to a representative reconstruction point, usually taken as the centroid of the cell. The proposed technique selects the reconstruction point within the quantization partition cell which results in a reconstructed image that best fits a non-Gaussian Markov Random Field image model. This approach results in a convex constrained optimization problem that can be solved iteratively. Efficient computational algorithms can be used in the minimization. This technique is extended to the post-processing of video sequences. The proposed approach provides a reconstructed image with reduced visibility of transform coding artifacts and superior perceived quality.

36

Acknowledgments

This work was funded partially by grant NL-213-456.

2.1 INTRODUCTION

Source coding of image data has been a very active area of research for many years. The goal is to reduce the number of bits needed to represent an image while making as few perceptible changes to the image as possible. Many algorithms have been developed which can successfully compress a grayscale image to around 0.8 bits per pixel (bpp) with almost no perceptible effects. A problem arises, however, as these compression techniques are pushed beyond this rate. For higher compression ratios ($<$ 0.4 bpp for grayscale) most algorithms start to generate artifacts which severely degrade the perceived quality of the image. The type of artifacts generated is dependent on the compression technique and on the particular image. For example, for motion-compensated block encoded video sequences, the most noticeable artifact is generally the discontinuities present at block boundaries due to both the transform coding of intraframe coded images and the motion compensation of interframe coded images.

This chapter proposes a technique which addresses this issue through a postprocessing algorithm that greatly reduces the artifacts introduced by the compression scheme. The technique is introduced for the processing of still images, but its extension to video sequences is straightforward. It is based on a stochastic framework in which probabilistic models are used for both the quantization noise introduced by the coding and for a "good" image. The restored image is the maximum a posteriori (MAP) estimate based on these models.

This chapter will first describe a generic model for image compression. For the purpose of the reconstruction algorithm, this model is general enough to describe many compression techniques. While other methods describe the characteristics of particular coding artifacts, the previously proposed image model [13], [12] used here incorporates prior knowledge about the properties of a good image. With the image model and the compression model, the MAP estimation problem becomes a constrained minimization problem. The gradient projection algorithm [1] which is used to solve this optimization problem is presented. The projection operator for scalar quantization is also described. The proposed techniques is extended to the postprocessing of video sequences. An implementation of the proposed technique that greatly reduces its computational load is introduced. Experimental results are shown for still images and video sequences. The postprocessing technique is applied on an image compressed using the JPEG [10] standard algorithm and on video sequences compressed following the H.261 [5] and H.263 [6] video coding standards. All these algorithms employ block-DCT transform followed by scalar quantization. It can be seen that images reconstructed with this new postprocessing technique show a reduction in many of the most noticeable artifacts.

2.2 DECOMPRESSION ALGORITHM

To decompress the proposed image representation, a MAP technique is proposed. Maximum a Posteriori (MAP) filters or estimators belong to a larger class of estimation schemes termed Bayesian estimation [7]. Let the compressed image data be represented by $\mathbf{y}$ while $\mathbf{z}$ is the decompressed full resolution image. For MAP estimation, the decompressed image estimate $\hat{\mathbf{z}}$ is given by

$$\hat{\mathbf{z}} = \arg\max_{\mathbf{z}} L(\mathbf{z}|\mathbf{y}), \tag{2.1}$$

where $L(\cdot)$ is the log likelihood function $L(\cdot) = \log Pr(\cdot)$, which in this case is the measure of how likely the decompressed image $\mathbf{z}$ resulted in the given compressed representation $\mathbf{y}$. Using Bayes' rule

$$
\begin{aligned}
\hat{\mathbf{z}} &= \arg\max_{\mathbf{z}} \left\{ \log \frac{Pr(\mathbf{y}|\mathbf{z})Pr(\mathbf{z})}{Pr(\mathbf{y})} \right\}, & (2.2) \\
&= \arg\max_{\mathbf{z}} \left\{ \log Pr(\mathbf{y}|\mathbf{z}) + \log Pr(\mathbf{z}) - \log Pr(\mathbf{y}) \right\}, & (2.3) \\
&= \arg\max_{\mathbf{z}} \left\{ \log Pr(\mathbf{y}|\mathbf{z}) + \log Pr(\mathbf{z}) \right\}, & (2.4)
\end{aligned}
$$

where the $Pr(\mathbf{y})$ term is dropped because it is a constant with respect to the optimization parameter $\mathbf{z}$. The conditional probability $Pr(\mathbf{y}|\mathbf{z})$ is based on the image compression method while the probability $Pr(\mathbf{z})$ is based on prior information about the image data.

2.2.1 Image Compression Model

In a transform coding compression technique, a unitary transformation H is applied to the original image $\mathbf{x}$. The compressed representation $\mathbf{y}$ is obtained by applying a quantization Q to the transform coefficients which can be written as

$$\mathbf{y} = Q[H\mathbf{x}]. \tag{2.5}$$

Quantization partitions the transform coefficient space and maps all points in a partition cell to a representative reconstruction point, usually taken as the centroid of the cell. The indices of these cells are usually entropy coded and then transmitted as the compressed representation $\mathbf{y}$. In the standard image decompression method, the reconstructed image is given by

$$\hat{\mathbf{z}} = H^{-1}Q^{-1}[\mathbf{y}], \tag{2.6}$$

where the inverse quantization maps the indices to the reconstruction points.

Since quantization is a many-to-one operation, many images map into the same compressed representation. The operation of the quantizer is assumed to be noise free;

that is, a given image $\mathbf{z}$ will be compressed to the same compressed representation $\mathbf{y}$ every time. The conditional probability for the noise free quantizer can be described by

$$Pr(\mathbf{y}|\mathbf{z}) = \begin{cases} 1, & \mathbf{y} = Q[H\mathbf{z}], \\ 0, & \mathbf{y} \neq Q[H\mathbf{z}]. \end{cases} \tag{2.7}$$

Since

$$\log Pr(\mathbf{y}|\mathbf{z}) = \begin{cases} 0, & \mathbf{y} = Q[H\mathbf{z}], \\ -\infty, & \mathbf{y} \neq Q[H\mathbf{z}]. \end{cases} \tag{2.8}$$

the MAP estimation in 2.4 can be written as the constrained optimization problem

$$\hat{\mathbf{z}} = \arg\min_{\mathbf{z} \in \mathcal{Z}} \left\{ -\log Pr(\mathbf{z}) \right\}, \tag{2.9}$$

where $\mathcal{Z}$ is the set of images which compress to $\mathbf{y}$, i.e. $\mathcal{Z} = \{\mathbf{z} : \mathbf{y} = Q[H\mathbf{z}]\}$.

2.2.2 Image Model

For a model of a "good" image, i.e. $Pr(\mathbf{z})$, a non-Gaussian Markov random field (MRF) model is used in [13] and [12]. This model has been shown to successfully model both the smooth regions and the discontinuities present in images.

A Gibbs distribution is used to explicitly write the distribution of MRF's. A Gibbs distribution is any distribution which can be expressed in the form

$$Pr(\mathbf{x}) = \frac{1}{Z} \exp\left\{ -\sum_{c \in \mathcal{C}} V_c(\mathbf{x}) \right\} \tag{2.10}$$

where Z is a normalizing constant, $V_c(\cdot)$ is any function of a local group of points c and $\mathcal{C}$ is the set of all such local groups. Note that the Gaussian model is a special case of the MRF. To understand how to include discontinuities into the statistical model, it is important to first understand what the model represents and how to define the model for a particular application. For a particular source signal $\mathbf{z}_1$, the value of the probability measure $Pr(\mathbf{z}_1)$ is related to how closely $\mathbf{z}_1$ matches our prior information about the source. So a $\mathbf{z}_1$ which closely matches our prior information should have a higher probability than one that does not. For this to be true, the function $V_c(\cdot)$ should provide a measure of the consistency of a particular $\mathbf{z}$, where a $\mathbf{z}$ more consistent with the prior information will have smaller values of $V_c(\cdot)$. The situation which is important in this chapter occurs when the prior information is mostly true but a limited amount of inconsistency is allowable (e.g., a piecewise smooth surface; that is, a surface which is mostly smooth but a few discontinuities are allowable).

In this chapter a special form of the MRF is used which has this very desirable property. This model is characterized by a special form of the Gibbs distribution

$$Pr(\mathbf{x}) = \frac{1}{Z} \exp\left\{ -\frac{1}{\lambda} \sum_{c \in \mathcal{C}} \rho_T(\mathbf{d}_c^t \mathbf{x}) \right\}, \tag{2.11}$$

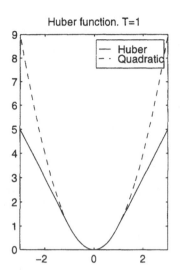

Figure 2:1   Huber minimax function $\rho_T(\cdot)$ superimposed on quadratic function.

where λ is a scalar constant that is greater than zero, d_c is a collection of linear operators and the function $\rho_T(\cdot)$ is given by

$$\rho_T(u) = \begin{cases} u^2, & |u| \leq T, \\ T^2 + 2T(|u| - T), & |u| > T; \end{cases} \qquad (2.12)$$

see Fig. 2.1. Since $\rho_T(\cdot)$ is convex, this particular form of the MRF results in a convex optimization problem when used in the MAP estimation formulation 2.4. Therefore, such MAP estimates will be unique, stable and can be computed efficiently. The function $\rho_T(\cdot)$ is known as the Huber minimax function and for that reason this statistical model is called the Huber-Markov random field model (HMRF).

For this distribution, the linear operators d_c provide the mechanism for incorporating what is considered consistent most of the time, while the function $\rho_T(\cdot)$ is the mechanism for allowing some inconsistency. The parameter T controls the amount of inconsistency allowable. The function $\rho_T(\cdot)$ allows some inconsistency by reducing the importance of the consistency measure when the value of the consistency measure exceeds some threshold, T.

For the measure of consistency, the fact that the difference between a pixel and its local neighbors should be small is used; that is, there should be little local variation in the image. For this assumption, an appropriate set of consistency measures is

$$\{d_c^t z\}_{c \in C} = \{z_{m,n} - z_{k,l}\}_{k,l \in \mathcal{N}_{m,n}}, \quad 1 \leq m, n \leq N, \qquad (2.13)$$

where $\mathcal{N}_{m,n}$ consists of the eight nearest neighbors of the pixel located at (m,n) and N is the dimension of the image. Across discontinuities this measure is large, but compared with a quadratic model, the relative importance of the measure at such a point is reduced because of the use of the Huber function.

The MAP estimate can now be written as

$$\hat{\mathbf{z}} \;=\; \arg\min_{\mathbf{z}\in\mathcal{Z}} \sum_{c\in\mathcal{C}} V_c(\mathbf{z}) \tag{2.14}$$

$$=\; \arg\min_{\mathbf{z}\in\mathcal{Z}} \sum_{1\le m,n\le N}\; \sum_{k,l\in\mathcal{N}_{m,n}} \rho_T(\mathbf{z}_{m,n}-\mathbf{z}_{k,l}). \tag{2.15}$$

As a result of the choice of image model, this results in a convex (but not quadratic) constrained optimization which can be solved using iterative techniques.

2.3 RECONSTRUCTION ALGORITHM

An iterative approach is used to find $\hat{\mathbf{z}}$ in the constrained minimization of 2.15. An initial estimate $\mathbf{z}^{(0)}$ is improved by successive iterations until the difference between $\mathbf{z}^{(k)}$ and $\mathbf{z}^{(k+1)}$ is below a given threshold ϵ. The rate of convergence of the iteration is affected by the choice of the initial estimate. A better initial estimate will result in faster convergence. The initial estimate used here is formed by the standard decompression

$$\mathbf{z}^{(0)} = H^{-1}Q^{-1}\left[\mathbf{y}\right]. \tag{2.16}$$

Given the estimate at the kth iteration, $\mathbf{z}^{(k)}$, the gradient projection method [1] is used to find the estimate at the next iteration, $\mathbf{z}^{(k+1)}$. The gradient of $\sum \rho_T\left(\mathbf{d}_c\mathbf{z}\right)$ is used to find the steepest direction $\mathbf{g}^{(k)}$ towards the minimum

$$\mathbf{g}^{(k)} = \nabla\left(\sum_{c\in\mathcal{C}} \rho_T\left(\mathbf{d}_c^t\mathbf{z}^{(k)}\right)\right) = \sum_{c\in\mathcal{C}} \rho_T'\left(\mathbf{d}_c^t\mathbf{z}^{(k)}\right)\mathbf{d}_c^t, \tag{2.17}$$

where $\rho_T'(u)$ is the first derivative of the Huber function. The size of the step $\alpha^{(k)}$ is chosen as

$$\alpha^{(k)} = \frac{\mathbf{g}^{(k)t}\mathbf{g}^{(k)}}{\mathbf{g}^{(k)t}\left(\sum_{c\in\mathcal{C}} \rho_T''\left(\mathbf{d}_c^t\mathbf{z}^{(k)}\right)\mathbf{d}_c\mathbf{d}_c^t\right)\mathbf{g}^{(k)}}. \tag{2.18}$$

This choice of step size is based on selecting the optimal step size for a quadratic approximation to the nonquadratic function in 2.15. Since this is an approximation, the value of the objective function may increase if the step size is too large. To avoid this potential problem, the value of $\alpha^{(k)}$ is divided by 2 until the step size is small enough that the value of the objective function is decreased.

Since the updated estimate $\mathbf{w}^{(k+1)}$,

$$\mathbf{w}^{(k+1)} = \mathbf{z}^{(k)} + \alpha^{(k)}\mathbf{g}^{(k)}, \tag{2.19}$$

may fall outside the constraint space $\mathcal{Z}$, $\mathbf{w}^{(k+1)}$ is projected onto $\mathcal{Z}$ to give the image estimate at the $(k+1)$-th iteration

$$\mathbf{z}^{(k+1)} = \mathcal{P}_{\mathcal{Z}}\left(\mathbf{w}^{(k+1)}\right). \tag{2.20}$$

In projecting the image $\mathbf{w}^{(k+1)}$ onto the constraint space $\mathcal{Z}$, we are finding the point $\mathbf{z}^{(k+1)} \in \mathcal{Z}$ for which $\|\mathbf{z}^{(k+1)} - \mathbf{w}^{(k+1)}\|$ is a minimum. If $\mathbf{w}^{(k+1)} \in \mathcal{Z}$, then $\mathbf{z}^{(k+1)} - \mathbf{w}^{(k+1)}$ and $\|\mathbf{z}^{(k+1)} - \mathbf{w}^{(k+1)}\| = 0$. Since H is unitary,

$$\|H\mathbf{z}^{(k+1)} - H\mathbf{w}^{(k+1)}\| = \|\mathbf{z}^{(k+1)} - \mathbf{w}^{(k+1)}\| \tag{2.21}$$

and the projection can be carried out in the transform domain.

The form of $\mathcal{P}_{\mathcal{Z}}$ is dependent on the quantizer Q. The projection operator $\mathcal{P}_{\mathcal{Z}}$ for scalar quantization is described in the following section. For the vector quantization case refer to [9].

2.3.1 Scalar Quantization Projection Operator

A scalar quantizer is a partition of the real number line $\mathcal{R}$ and is defined by its breakpoints. The breakpoints are determined by the amount of compression desired. The boundaries or breakpoints of cell i are $l[i]$ and $h[i]$. If a coefficient ω falls in cell i,

$$l[i] \leq \omega \leq h[i], \tag{2.22}$$

then inx i is transmitted. The compressed representation $\mathbf{y}$ is the set of indices for all the transform coefficients.

Projection to the constraint space is rather simple for scalar quantization. To find the transform coefficients μ in $H\mathbf{z}^{(k+1)}$, find the corresponding coefficient ω in $H\mathbf{w}^{(k+1)}$. From the compressed representation $\mathbf{y}$, find in which cell μ should fall. Assume μ should fall in cell i. The standard decompression method would use $\mu = \tau$ where τ is the centroid of the cell. The coefficient μ which falls in cell i can be determined by

$$\mu = \begin{cases} l[i], & \omega < l[i], \\ \omega, & l[i] \leq h[i], \\ h[i], & h[i] \leq \omega. \end{cases} \tag{2.23}$$

See Fig. 2.2 for an example of the projection operator. This will minimize $\|H\mathbf{z}^{(k+1)} - H\mathbf{w}^{(k+1)}\|$ while assuring that $\mathbf{y} = Q\left[H\mathbf{z}^{(k+1)}\right]$. The image estimate $\mathbf{z}^{(k+1)}$ is obtained by performing the inverse transformation on $H\mathbf{z}^{(k+1)}$.

2.3.2 Postprocessing of Video Sequences

All standard compression algorithms provide a compressed video sequence that is a combination of intra and interframe pictures. Moreover, there can be two types of

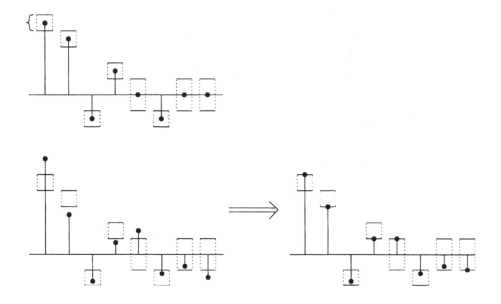

Figure 2.2 Projection operator; (top) original coefficients, represented by dots inside of the allowable range; (bottom left) coefficients of the updated image $\mathbf{w}^{(k+1)}$; (bottom right) coefficients after projection

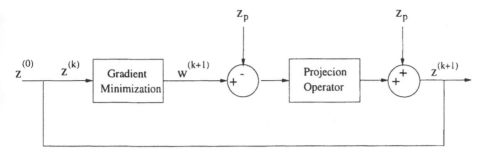

Figure 2.3   MAP postprocessing for interframe coded images.

interframe encoded pictures: forward predicted and bidirectionally predicted pictures. The implementation of the algorithm, as it was presented in the previous section, is immediate in the case of intraframe coding.

In interframe pictures the reconstructed pels values are the result of the addition of two parts: a predicted term and an error residual. In the literature this last term is usually called Displaced Frame Difference (DFD). Therefore the decompressed image $z^{(0)}$ can be written,

$$z^{(0)} = z_p + z_{DFD}, \tag{2.24}$$

where z_p is the predicted part using motion vector information and z_{DFD} is the error residual recovered from the compressed representation. In forward prediction mode, the predicted term is the value of the pixel of the previous frame pointed to by the motion vectors. In bidirectional mode, the predicted part is a linear combination of pel values corresponding to past and future frames pointed to by forward and backward motion vectors, respectively. Assuming that the compressed pictures are used in the prediction loop of the encoder and because the motion vectors are transmitted lossless, the predicted term at the decoder is exactly the same as it was at the encoder. This assumption is valid for all the standard algorithms.

In terms of the MAP decoding algorithm, the only difference with the case of still images is that the projection operator is applied over the DCT coefficients of the error residual z_{DFD}. Fig. 2.3 shows the decoding structure for interframe coded images. Notice that at each iteration, the predicted term z_p should be removed from the reconstructed image before the projection operator is applied. The predicted term z_p is added back after the projection into the constraint set.

With some appropriate modifications on the constraint set $\mathcal{Z}$, the subtraction and addition of z_p can be avoided [8]. Based on equation 2.24 the constraint set $\mathcal{Z}$ is defined as

$$\mathcal{Z} = \left\{ \mathbf{z} : Z_{u,v}^{\min} \leq H[(\mathbf{z} - \mathbf{z_p})]_{u,v} \leq Z_{u,v}^{\max}, 1 \leq u, v \leq N \right\}, \tag{2.25}$$

where $Z_{u,v}^{min}$ and $Z_{u,v}^{max}$ determine the quantization interval of the $\{u, v\}$-th DFD coefficient. Assuming that the received coefficient $Z_{DFD_{u,v}}$ lies in the center of the quantization cell we can write

$$\begin{aligned} Z_{u,v}^{\min} &= Z_{DFD_{u,v}} - \tfrac{1}{2} Q_{u,v} \\ Z_{u,v}^{\max} &= Z_{DFD_{u,v}} + \tfrac{1}{2} Q_{u,v} \end{aligned}, \tag{2.26}$$

where $Q_{u,v}$ is the quantizer step size for the $\{u, v\}$-th DFD coefficient.

Moreover,

$$\begin{aligned} Z_{u,v}^{\min} &= H[\mathbf{z}_{DFD}]_{u,v} - \tfrac{1}{2} Q_{u,v} \\ Z_{u,v}^{\max} &= H[\mathbf{z}_{DFD}]_{u,v} + \tfrac{1}{2} Q_{u,v} \end{aligned}. \tag{2.27}$$

Combining 2.25 and 2.27 and remembering the linearity of the transformation H, the constraint set $\mathcal{Z}$ can be written as

$$\mathcal{Z} =$$
$$\left\{ \mathbf{z} : H[\mathbf{z}_{DFD}]_{u,v} - \frac{1}{2} Q_{u,v} \leq H[\mathbf{z}]_{u,v} - H[\mathbf{z_p}]_{u,v} \leq H[\mathbf{z}_{DFD}]_{u,v} + \frac{1}{2} Q_{u,v}, \right.$$
$$\left. 1 \leq u, v \leq N \right\}. \tag{2.28}$$

By adding the term $H[\mathbf{z}_p]_{u,v}$ to both sides of both inequalities and applying again the linearity of the transformation,

$$\mathcal{Z} =$$
$$\left\{ \mathbf{z} : H[\mathbf{z}_p + \mathbf{z}_{DFD}]_{u,v} - \frac{1}{2} Q_{u,v} \leq H[\mathbf{z}]_{u,v} \leq H[\mathbf{z}_p + \mathbf{z}_{DFD}]_{u,v} + \frac{1}{2} Q_{u,v}, \right.$$
$$\left. 1 \leq u, v \leq N \right\} =$$
$$\left\{ \mathbf{z} : H[\mathbf{z}^{(0)}]_{u,v} - \frac{1}{2} Q_{u,v} \leq H[\mathbf{z}]_{u,v} \leq H[\mathbf{z}^{(0)}]_{u,v} + \frac{1}{2} Q_{u,v}, \right.$$
$$\left. 1 \leq u, v \leq N \right\}. \tag{2.29}$$

Therefore the projection operator $\mathcal{P}_Z$ is given by:

$$\mathbf{z} = \mathcal{P}_Z(\mathbf{w}) = H^{-1}[\mathbf{Z}], \tag{2.30}$$

where

$$\mathbf{Z} = \{ Z_{u,v} : 1 \leq u, v \leq N \} \tag{2.31}$$

and

$$
Z_{u,v} = \begin{cases}
H[\mathbf{z}^{(0)}]_{u,v} - \frac{1}{2}Q_{u,v} & \text{if } H[\mathbf{w}]_{u,v} < H[\mathbf{z}^{(0)}]_{u,v} - \frac{1}{2}Q_{u,v} \\
H[\mathbf{z}^{(0)}]_{u,v} + \frac{1}{2}Q_{u,v} & \text{if } H[\mathbf{w}]_{u,v} > H[\mathbf{z}^{(0)}]_{u,v} + \frac{1}{2}Q_{u,v} \\
H[\mathbf{w}]_{u,v} & \text{if } H[\mathbf{z}^{(0)}]_{u,v} - \frac{1}{2}Q_{u,v} \leq H[\mathbf{w}]_{u,v} \leq H[\mathbf{z}^{(0)}]_{u,v} + \frac{1}{2}Q_{u,v}
\end{cases}
$$

$$(2.32)$$

This implies that the position of every cell, in which the coefficients $H[\mathbf{w}]_{u,v}$ should lie, is given by the coefficients of the image obtained by standard decompression and the amplitude of these cells is identical to the step size of the quantizer used to quantize the error residual. Compared to the case of intraframes pictures the only difference is that the bounds of the projection operator will depend on the quantization steps of the error residual instead of the ones from the pel values.

2.4 IMPLEMENTATION ISSUES

The main difficulty regarding implementation of the proposed MAP filter lies in its computational complexity. First, the gradient of the objective function must be evaluated. A proper step size must then be determined, by evaluating equation 2.18. After scaling the gradient and subtracting the result from the image, the objective function must be evaluated to ensure that it was in fact decreased. If it was not actually decreased, the step size must be successively halved until the objective function is decreased. The projection must then be performed. These operations are all very expensive from a computational standpoint, and methods will be presented in this section that allow these operations to be performed in a more feasible manner.

2.4.1 The Step Size

Rather than using the gradient of the objective function for the entire image (as presented previously), a different approach will be taken for better efficiency. Here, only one pixel will be considered at a time, rather than the entire image. Also, no second-order derivatives will be used in the determination of the change to be applied, as was done in equation 2.18. On the contrary, a *constant* step size will be multiplied by the gradient of the objective function of the single pixel being considered. Each pixel will be modified in the image, after which the projection will be applied.

The derivation of this constant step size will be conducted in terms of a single pixel and its eight neighbors, and then the concept will be extended to include results for the entire image. Let the pixel under consideration be x, and let its eight nearest neighbors be n_i, $i = 1, \ldots, 8$. Assume for the following discussion that in evaluating the cost function, L of the eight neighbors are used, and n_i, $i = 1, \ldots, L$ are those L neighbors.

The cost function for this single pixel is

$$R(x) = \sum_{i=1}^{L} \rho_T(x - n_i), \qquad (2.33)$$

and following a change applied to x of size Δx, the cost function becomes

$$R_\Delta(x) = \sum_{i=1}^{L} \rho_T(x - \Delta x - n_i). \qquad (2.34)$$

If it is assumed that the final value of Δx will be relatively small, then 2.34 can be approximated using a Taylor series expansion by

$$\hat{R}_\Delta(x) = \sum_{i=1}^{L} \left\{ \rho_T(x - n_i) - \Delta x \rho_T'(x - n_i) \right\}. \qquad (2.35)$$

The Δx that minimizes this can be found by taking the derivative and setting it equal to zero,

$$\sum_{i=1}^{L} \left\{ \rho_T'(x - n_i) - \Delta x \rho_T''(x - n_i) \right\} = 0, \qquad (2.36)$$

which gives the optimal Δx as

$$\Delta x_{opt} = \frac{\sum_{i=1}^{L} \rho_T'(x - n_i)}{\sum_{i=1}^{L} \rho_T''(x - n_i)}. \qquad (2.37)$$

This equation is of the form (step size) $\times$ (gradient), where the numerator is the gradient of the objective function for the pixel x, and the step size is the reciprocal of the denominator. Since the goal here is to have better computational efficiency, the step size will be replaced with a constant, hence avoiding the second derivatives present in 2.37. The approach taken here is to find the *minimum* value that the step size may assume (and hence the largest value that the denominator in 2.37 may assume), and use that as the constant step size. The motivation for finding a small step size is to try to avoid overshooting whatever the pixel value is that actually minimizes the cost function for the pixel x, which could possibly result in increasing the cost function.

The minimum step size is then given by

$$\alpha_{min} = \min \left\{ \frac{1}{\sum_{i=1}^{L} \rho_T''(x - n_i)} \right\}, \qquad (2.38)$$

$$= \frac{1}{L \max_u \rho_T''(u)}, \qquad (2.39)$$

$$= \frac{1}{2L}, \qquad (2.40)$$

where $\max_u \rho_T''(u) = 2$, and $\alpha_{min} = \frac{1}{2L}$ is ultimately the quantity of interest. Thus the change applied to x will be

$$\Delta x = \frac{1}{2L} \sum_{i=1}^{L} \rho_T'(x - n_i).$$ (2.41)

The previous development was based on the approximation in 2.35. An important question is whether or not using that approximation actually leads to a step size that is guaranteed to decrease the cost function. For example, if the approximation is not very good, the derived change may end up overshooting the minimum and actually increase the cost function.

To be more precise, it must be shown than the cost function is guaranteed to be decreased when the derived change from above is applied,

$$\sum_{i=1}^{L} \rho_T \left(x - n_i - \frac{1}{2L} \sum_{i=1}^{L} \rho_T'(x - n_i) \right) \overset{?}{\underset{,}{\leq}} \sum_{i=1}^{L} \rho_T \left(x - n_i \right).$$ (2.42)

In showing that 2.42 is true, it will suffice to show that when the change is applied to x, the resulting pixel x_{new} has not overshot whatever pixel minimizes the cost function. This is true because, since the move is in the direction of the negative derivative of the cost function, an arbitrarily small step will definitely decrease the cost function. Since the cost function is convex, as long as the new pixel does not go beyond the pixel that minimizes the cost function, the cost function cannot increase.

Now to prove the validity of 2.42, let x_R^* be the largest pixel value that minimizes the objective function; similarly, let x_L^* be the smallest pixel value that minimizes the objective function. Fig. 2.4 shows a hypothetical configuration of x_L^*, x_R^*, and x. Note that it is possible but not necessary that $x_L^* = x_R^*$. Without loss of generality, it will be assumed here that $x > x_R^*$. Due to the convex nature of the problem, if x is between x_L^* and x_R^*, the cost function is still minimized. For $x < x_L^*$ an argument analogous to the following can be applied.

For $x > x_R^*$ it must be shown that after the change has been applied to x, the result will not be less than x_R^*. If x does become less than x_R^*, there is the possibility that the cost function will increase. In particular, if the new x becomes less than x_L^*, then the cost function may increase.

The first important fact to note is that for $x = x_R^*$, the quantity Δx will be zero, because x_R^* minimizes the cost function, and hence

$$\sum_{i=1}^{L} \rho_T'(x_R^* - n_i) = 0.$$ (2.43)

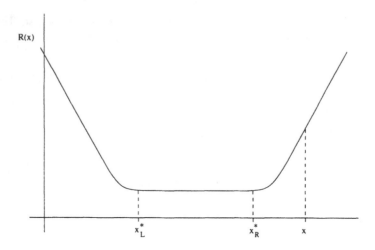

Figure 2.4   Hypothetical configuration of x_L^*, x_R^*, and x

Now, consider what happens as one moves slightly higher than x_R^*: Δx *must* be increasing linearly. That it must be increasing is true because, by definition, x_R^* is the highest pixel value such that $\Delta x = 0$. As x becomes larger than x_R^*, the cost function is increasing, which means that its derivative is positive, which in turn means Δx is positive. That it must be increasing linearly (piecewise linearly, to be more precise) is due to the nature of $\rho_T'(\cdot)$; see figure 2.5. The approach now taken to prove 2.42 will be to determine bounds for the size of Δx, and use these bounds to show that the new filtered pixel, x_{new}, will not overshoot x_R^*.

For x just slightly larger than x_R^*, the fastest that Δx may be increasing is when each of the L $\frac{1}{2L}\rho_T'(x - n_i)$ terms are increasing with a slope of $\frac{1}{L}$, which will yield a maximum slope of 1. The slowest that Δx may be increasing is when only one of the L $\frac{1}{2L}\rho_T'(x - n_i)$ terms is increasing with a slope of $\frac{1}{L}$, which yields a minimum slope of $\frac{1}{L}$. These could also have been determined by looking at the derivative of Δx with respect to x,

$$\frac{d}{dx}\Delta x = \frac{1}{2L}\sum_{i=1}^{L} \rho_T''(x - n_i), \tag{2.44}$$

which just represents the slope of Δx as a function of x. Since 2.44 must be greater than zero for x slightly higher than x_R^* (because it has already been determined that Δx is increasing for x slightly above x_R^*), the smallest slope Δx will be increasing with is when only one of the terms in 2.44 is 2; this yields a minimum slope of $\frac{1}{L}$. Similarly, the highest the slope can be is when all L terms in 2.44 are 2; this yields a maximum slope of 1.

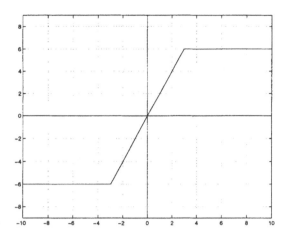

Figure 2.5   The first derivative of the Huber function, $\rho'_T(u), T = 3$.

So for x just larger than x^*_R, the maximum slope is 1, and the minimum slope is $\frac{1}{L}$. It must now be determined how far beyond x^*_R that these rates of increase can continue. This is simple enough: since the absolute largest value that Δx can assume is T (when each of the $L \frac{1}{2L} \rho'_T(x - n_i)$ terms is $\frac{T}{L}$), and if the maximum slope of 1 is assumed, the earliest that Δx can be T is at $x = x^*_R + T$. Similarly, at $x = x^*_R + T$ the minimum value Δx can be is $\frac{T}{L}$. These bounds can now be stated in a more precise nature. The bound on Δx from above can be written as

$$\frac{1}{2L} \sum_{i=1}^{L} \rho'_T(x - n_i) \leq \begin{cases} x - x^*_R, & x^*_R \leq x \leq x^*_R + T \\ T, & x > x^*_R + T \end{cases}. \qquad (2.45)$$

Similarly, the bound from below is

$$\frac{1}{2L} \sum_{i=1}^{L} \rho'_T(x - n_i) \geq \begin{cases} \frac{x - x^*_R}{L}, & x^*_R \leq x \leq x^*_R + T \\ \frac{T}{L}, & x > x^*_R + T \end{cases}. \qquad (2.46)$$

These are shown in Fig. 2.6.

There are thus two situations that must be examined: condition (a), where $x > x^*_R + T$; and condition (b), where $x^*_R \leq x \leq x^*_R + T$. Case (a) is simple. Since the change Δx must be between $\frac{T}{L}$ and T, the cost function will be decreased, with x_{new} being closer to x^*_R by at least $\frac{T}{L}$. Since it will not move by more than T, x_{new} will never overshoot x^*_R for $x > x^*_R + T$.

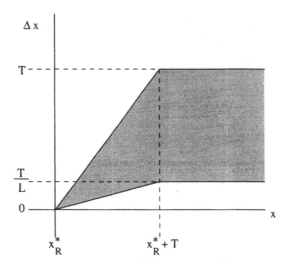

Figure 2.6 Upper and lower bounds for Δx_{min}.

For case (b), it has already been shown in 2.45 that $\Delta x \leq x - x_R^*$ from which it immediately follows that

$$x - \Delta x \quad \geq \quad x_R^*, \qquad (2.47)$$

$$x_{new} \quad \geq \quad x_R^*. \qquad (2.48)$$

Thus one sees that x_{new} will never overshoot the optimal value x_R^*, and thus will never increase the cost function.

As mentioned previously, the above developments can be applied analogously for $x < x_L^*$, with corresponding results.

The extension of the above developments to include the entire image is straightforward. It is proposed here that the methods presented previously for a single pixel merely be applied to each pixel in the image. When a pixel value is changed, the modified pixel will be used when filtering subsequent pixels; thus this method is recursive in nature. For example, one may filter an image by filtering across rows, as shown in Fig. 2.7. After a single iteration on a particular pixel, the algorithm moves on to the next pixel to be filtered. When using this recursive pixel-by-pixel method, the term *iteration* will refer to an iteration on each pixel in the image, as opposed to an iteration on a single pixel within an image.

When using this method for the entire image, the objective function for the entire image is guaranteed to decrease, or at least not increase if the minimum is already attained. This is because when a pixel x is filtered, the only terms in the total objective

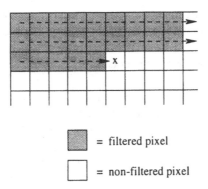

　　　　　■ = filtered pixel

　　　　　□ = non-filtered pixel

Figure 2.7 Example of filtering using recursive pixel-by-pixel approach. Note the recursive nature of this method: when the pixel x is being changed, both filtered and non-filtered pixels are being used in the computation of Δx.

function that will change are those terms involving x,

$$R(x) = \sum_{i=1}^{L} \rho_T(x - n_i). \tag{2.49}$$

However, it has already been shown that this sum will never increase when using the method outlined previously. Thus as each pixel is filtered, the cost function is non-increasing, and when the entire image has been filtered, the objective function will be less than or equal to the objective function prior to that particular iteration.

There are several advantages of using the constant step size method derived in this section. First, since it is known that the objective function will decrease, the objective function need not ever be evaluated. Second, when determining how much to modify a pixel by, there are no costly multiply/divide operations necessary; a look at 2.41 reveals that only adds/subtracts, compares, and shifts are necessary, assuming that L is an even power of 2 (which it usually will be).

2.4.2 The Projection

Due to its widespread use in image and video compression, the block discrete cosine transform (BDCT) will be the assumed image transformation.

As explained previously, the projection involves performing the forward transform, clipping the resulting coefficient values to the range dictated by the quantizer, and then inverse transforming to yield the projected image. If 8 by 8 blocks are used, then 64 coefficients will be computed and clipped to their allowable ranges. It is proposed here that, especially for low bit rates, all 64 coefficients need *not* be computed and clipped. On the contrary, only coefficients that lie within a sub-block of the original block will be computed and clipped. This sub-block will be defined as the smallest square sub-block, with one corner anchored on the DC component, that contains all of the non-zero transform coefficients. Several examples of such sub-blocks are shown in Fig. 2.8.

The reasoning behind using this "partial" projection is as follows: the non-zero transform-domain coefficients must certainly be clipped, because they represent important information for the image that should not be removed by the MAP smoothing operation. However, considering that MAP filtering is essentially a low-pass operation, and since zero-valued transform coefficients typically occur at high frequencies, it would be less likely for these zero-valued high-frequency coefficients to be removed from their allowable range by MAP filtering.

Another reason for not constraining these previously-zero transform coefficients is due to the nature of quantization. Zero-valued coefficients are usually encouraged by some means, especially at higher spatial frequencies. For example, JPEG baseline usually allocates less bits for higher spatial frequencies. Similarly, quantization such as in H.263 uses a central dead zone about zero, which essentially means that a larger range of values will be mapped to zero than will be mapped to a particular non-zero number. The implications of this are as follows: The MAP procedure will probably move many of these zero-valued coefficients from zero. However, due to the quantization, these coefficients can move by a larger amount without needing to be clipped than could other coefficients that were previously non-zero.

It has been shown [11] that only clipping the coefficients in the smallest sub-square of non-zero coefficients, compared to performing the actual projection, has no noticeable effects from a subjective viewpoint, and insignificant effects from an objective viewpoint.

The main motivation for using the partial constraints is to decrease the computational requirements of MAP filtering. For DCT coding of 8 by 8 blocks, 16 one-dimensional DCT's of length 8 need be evaluated each time the projection is applied, as well as the corresponding 16 one-dimensional IDCT's after clipping has been performed. Using the partial constraints, the computations can be significantly reduced. Let S_0 be the size of the sub-block to be used for a particular block. Then $(8 + S_0)$ one-dimensional DCT's that *only need to compute S_0 of the 8 transform coefficients* must be performed. Similarly, $(8 + S_0)$ one-dimensional IDCT's that only take S_0 of the 8 transform coefficients and compute 8 spatial-domain values need to be performed. This operation is shown for a partial DCT of size 3 ($S_0 = 3$) in Fig. 2.9.

The computational cost of computing partial DCT/IDCT's of size S_0 can be significantly less than computing the complete DCT/IDCT's. When computing S_0 of the 8 transform coefficients, many of the operations necessary for the full transform can be eliminated. An example demonstrating this is shown in Fig. 2.10 for a partial DCT of size 3; the unnecessary operations have been de-emphasized. In the figure, b0–b7 are

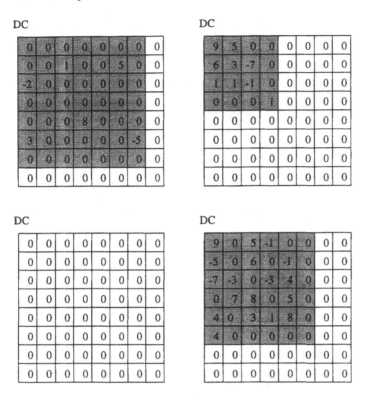

Figure 2.8   Original transform-domain coefficient blocks, and their sub-squares of sizes 7, 4, 0, and 6.

the spatial-domain values, and d0–d2 are the three lowest-frequency transform-domain values returned by the partial DCT of size 3. As can be seen, 7 additions and 9 multiplications have been eliminated from the original 26 additions and 16 multiplications. However, since only $8 + S_0 = 11$ 1-D partial DCT's need to be evaluated (instead of 16 full DCT's), the savings will be higher. For a full 8 by 8 DCT, $26 \cdot 16 = 416$ additions and $16 \cdot 16 = 256$ multiplications need to be evaluated. For a 3 by 3 partial DCT, only $19 \cdot (8 + 3) = 209$ additions and $7 \cdot (8 + 3) = 77$ multiplications need to be evaluated.

One of course may wonder what overall computational gains can be achieved by using this sub-square method as opposed to using the full projection. For medium- to high-quality images there is little gain, for sub-square sizes are not significantly smaller than 8 very often, and most of the coefficients are being constrained. However, for low-quality images, and low to moderate bit rates of digital video, significant gains can

54

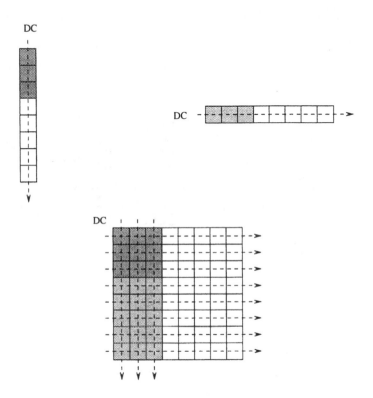

Figure 2.9   Evaluation of partial DCT of size 3: 1-D partial DCT of size 3 on column; 1-D partial DCT of size 3 on row; and 2-D partial DCT of size 3 on block.

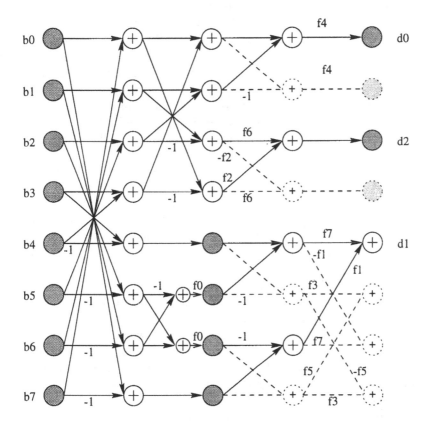

Figure 2.10   One-dimensional partial DCT of size 3, with unnecessary nodes de-emphasized

be achieved. This is naturally due to the increased number of small sub-block sizes; this is especially true for video, where block residuals (after motion compensation) typically have very little high-frequency energy.

2.4.3 A Look at the MAP Parameters

In an actual implementation of the MAP algorithm, the procedures described in the previous sections should certainly be used. These procedures provide more efficient methods of performing the optimization involved in the MAP algorithm than the other methods presented. However, there are still important decisions to make in an actual filtering situation. In particular, there are three degrees of freedom for the MAP filter—the parameter T in the Huber function; the number of neighbors used, L; and the number of iterations. These three parameters essentially determine what the quality of the filtered image will be. Although not discussed in detail here, the number of iterations and the number of neighbors also have a direct impact on the algorithm's computational complexity.

The parameter T of the Huber function determines the amount of smoothing that will be done to an image. Larger values of T result in more smoothing, while smaller values of T result in less smoothing. From a theoretic point of view, using a large value for T means that the image assumes, in a probabilistic sense, that pixels in a neighborhood will be very close in value. Similarly, a small value of T implies that the image model will allow more variation between pixels in a neighborhood. Experimentally, using values of T between 1 and 10 appears to provide satisfactory results. The experiments discussed in this chapter were typically conducted with T equal to 1, 2, or 3. A decision for T should be based upon the types of images expected for a particular application. For example, video teleconferencing applications would probably not require the same T that music video applications would.

For all experiments discussed in this paper, the number of neighbors used was 8, i.e., all of the nearest neighbors for each pixel were used. One may also use $L = 4$, which would involve using only the horizontal and vertical neighbors. Using $L = 4$ does reduce the computational complexity of the algorithm, but the results do not look as good as when all of the eight neighbors are used. This should come as no surprise, because using $L = 4$ does not take advantage of all the information that is available in a pixel's neighborhood. However, for block transform coding, blocking artifacts only occur in the horizontal and vertical directions. Thus using just the horizontal and vertical neighbors for the MAP procedure can satisfactorily reduce these blocking effects. However, for more general compression artifacts (e.g., ringing), using $L = 8$ is the best choice.

The choice for when to terminate the MAP algorithm is more subtle than the other MAP decisions. Strictly speaking, the true MAP estimate of the image will be the result once the MAP algorithm has converged to an image that minimizes the objective function and lies within the constraint space. From a practical standpoint, true convergence may take exceedingly long, and so some other criterion for stopping the algorithm must be adopted. There are any number of choices for terminating an optimization problem; for example, one may take some norm of the difference between images at consecutive iterations, and if this number is below some threshold the algorithm is terminated. However, this would be computationally expensive, and there is also the question of how appropriate typically-used norms are for images interpreted by the human visual system.

In practice, the best solution seems to be to terminate the algorithm after some predetermined number of iterations has been reached. The predetermined number of iterations would be chosen based on several factors: the severity of the expected compression artifacts; the computational resources one has available; and the desired "closeness" to actual convergence of the algorithm. Of these three, the main restriction is the available computing resources. If a single image is being filtered, and there are no strict time limitations, one may use more iterations to get the better estimate of the image. However, if there are strict time or computing limitations (a real time or near-real time video application, for example), then by necessity one is limited to fewer iterations.

Experimentally, the MAP algorithm seems to converge within 10 iterations. By "seems to converge", it is meant that no noticeable difference can be distinguished with further iterations. Thus if there are not strict limitations for computing time, 10 iterations will typically be sufficient. For situations where there are computational limitations, 2 to 5 iterations may have to suffice. In these situations, the main goal of MAP filtering would be to eliminate visually-annoying compression artifacts, and convergence is not as important. Typically 2 to 5 iterations will sufficiently reduce these artifacts. Two iterations would be considered as a lower bound, where the artifacts are not severe, and up to 5 iterations may be necessary for severe artifacts.

2.5 EXPERIMENTAL RESULTS

In this section, the results of using the proposed postprocessing method are shown for still images compressed using the JPEG standard algorithm [10] and the H.261 [5] and H.263 [6] video compression algorithms.

2.5.1 Still Images

The well-known "Lenna" image shown in Fig. 2.11 was compressed to 0.264 bpp by the JPEG algorithm, using the quantization table shown in Table 2.1.   The result of standard decompression of this image is shown in Fig. 2.12(top) and enlarged to show detail in Fig. 2.13(top).   At this compression ratio, the coding artifacts are very noticeable. Most noticeable are the blocking effects of the coding algorithm, but aliasing artifacts along large contrast edges are also noticeable. The result of

58

Figure 2.11   Lenna image: original (top), enlarged (bottom).

Figure 2.12   JPEG compressed images. Standard decompression (top), after 9th iteration of postprocessing algorithm $T = 1.0$ (bottom).

Figure 2.13 JPEG compressed images, enlarged view. Standard decompression (top), after 9th iteration of postprocessing algorithm $T = 1.0$ (bottom).

80	55	50	80	120	200	255	305
60	60	70	95	130	290	300	275
70	65	80	120	200	285	345	280
70	85	110	145	255	435	400	310
90	110	185	280	340	545	515	385
120	175	275	320	405	520	565	460
245	320	390	435	515	605	600	505
360	460	475	490	560	500	515	495

Table 2.1   JPEG quantization table used for Lenna image.

postprocessing this image is shown in Fig. 2.12 (bottom) and enlarged to show detail in Fig. 2.13 (bottom). The number of iterations needed to sufficiently reduce the artifacts is dependent on the severity of the degradation. More iterations are required for convergence with more severe degradation. Notice that the blocking effects have been removed in the postprocessed image. This can be most easily seen in the shoulder and background regions. Notice that while the discontinuities due to the blocking effects have been smoothed, the sharp discontinuities in the original image, such as along the hat brim, have been preserved. The full size image is 512×512 and the enlargements show the center 256×256 from the full size image.

Notice that high frequency information, such as the details in the hat of the original image, is completely lost when the DCT coefficients carrying this information are lost due to quantization. The postprocessing cannot reinvent this information which is totally absent from the compressed representation. However, for structures such as edges which have information content spread across low frequencies as well as high frequencies, the postprocessor is able to restore some high frequency information based on the low frequency components in the compressed representation. This allows crisper edges which contain high frequencies to appear in the restored image. See for example the hat brim and along the edge of the shoulder.

As introduced in previous sections, the MAP postprocessing algorithm can be easily applied on motion-compensated block-DCT compressed video sequences. H.261 [5] and H.263 [6] are the video communications standards sponsored by the International Telecommunication Union. Both of them are motion-compensated block-DCT compression algorithms.

Fig 2.14 shows a frame of the well-known "mother and daughter" QCIF sequence compressed at 60 Kbit/sec (15frames/sec) following the H.261 standard. The H.261 software codec [3] has been employed to obtain the results presented in this section. The decoder has been appropriately modified to implement the proposed postprocessing

62

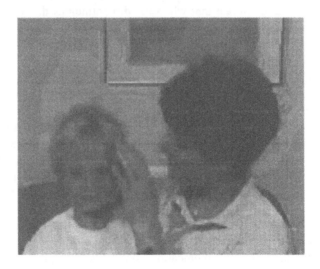

Figure 2.14   H.261 compressed sequence. Standard decompression (top), after 5th iteration of postprocessing algorithm, $T = 2.0$ (bottom).

algorithm. The same compressed sequence has been decoded using the standard decompression procedure and using the proposed technique. Even though, at this rate, the compression artifacts, particularly blocking effects, are quite severe, only five iterations are needed to significantly reduced the artifacts. In comparison with the still image case presented above, a larger value of T in the image model is employed since, due to the more severe blocking effects, a smaller degree of inconsistency is allowed. With this increment of T, the image model emphasizes the smooth surfaces divided by strong edges.

H.263 is an evolution of H.261 in the sense that they share the same basic motion-compensated block-DCT scheme, but H.263 incorporates some novel features that make it more efficient. See for example [2] for a comparison of both standards. Figures 2.15 to 2.18 show the gain in visual quality obtained in an H.263 compressed sequence thanks to the introduction on the MAP postprocessing algorithm. The famous "foreman" sequence has been employed for the tests. All the advanced H.263 negotiable options have been turned off. The H.263 software codec available at [4] has been used for the trials. Notice that only three iterations are needed to get significant improvements.

All video sequences used in this section are color sequences. Because of their similarities with natural images, the chrominance components, U and V, received the same postprocessing treatment as the luminance component Y.

2.6 CONCLUSION

The problem of image decompression has been cast as an ill-posed inverse problem, and a stochastic regularization technique has been used to form a well-posed reconstruction algorithm. A statistical model for the image was produced which incorporated the convex Huber minimax function. The use of the Huber minimax function $\rho_T(\cdot)$ helps to maintain the discontinuities from the original image which produces high resolution edge boundaries. Since $\rho_T(\cdot)$ is convex, the resulting multidimensional minimization problem is a constrained convex optimization problem. Particular emphasis has been given to the search for efficient computational algorithms to solve the minimization. Originally developed for the decompression of still images, the algorithm has been properly extended to be applied to the decompression of motion-compensated compressed video sequences. Nevertheless, the problem formulation and the solution are general enough that the proposed technique can be easily extended to other compression schemes such as subband decomposition and vector quantization. After extensive testing on still images and video sequences, the proposed algorithm has been shown to produce reconstructed images with much less noticeable coding artifacts than the images obtained using standard decompression techniques.

Figure 2.15   H.263 compressed sequence at 34.8 Kbit/sec and 10 frames/sec. Standard decompression (top), after 3rd iteration of the postprocessing algorithm, $T = 3.0$ (bottom).

Figure 2.16   Enlarged view of an H.263 compressed sequence at 34.8 Kbit/sec and 10 frames/sec. Standard decompression (top), after 3rd iteration of the postprocessing algorithm, $T = 3.0$ (bottom).

Figure 2.17 H.263 compressed sequence at 75.1 Kbit/sec and 10 frames/sec. Standard decompression (top), after 3rd iteration of the postprocessing algorithm, $T = 3.0$ (bottom).

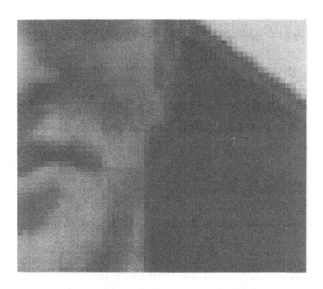

Figure 2.18 - Enlarged view of an H.263 compressed sequence at 75.1 Kbit/sec and 10 frames/sec. Standard decompression (top), after 3rd iteration of the postprocessing algorithm, $T = 3.0$ (bottom).

68

References

[1] M. S. Bazaraa, H. D. Sherali and C. M.Shetty, "Nonlinear Programming: Theory and algorithms", 2nd ed., Wiley, New York, 1993.

[2] V. Bhaskaran and K. Konstantinides, "Image and Video Compression Standards", 2nd ed., Kluwer Academic Publishers, Boston, 1997.

[3] H.261 Software Codec available at ftp.havefun.stanford.edu.

[4] H.263 Software Codec available at http://www.nta.no/brukere/DVC/

[5] International Telecommunication Union, ITU-T, "H.261. Video Codec for Audiovisual Services at $p \times 64$ kbits", 1993.

[6] International Telecommunication Union, ITU-T, "Draft H.263. Video Coding for Low Bitrate Communication", 1995.

[7] S. M. Kay, "Statistical Signal Processing", Prentice-Hall, Upper Saddle River, NJ, 1993.

[8] R. Llados-Bernaus and R. L. Stevenson, "Reduction of Coding Artifacts in Video Compression", in Proceedings of the SPIE Conf. on Electronic Imaging and Multimedia Systems, vol. 2898, pp. 2-10, Beijing (China), November 1996.

[9] T. P. O'Rourke and R. L. Stevenson, "Improved Image Decompression for Reduced Transform Coding Artifacts", IEEE Trans. Circ. Syst. Video Tech., vol. 5, no. 6, pp. 490-499, Dec. 1995.

[10] W. B. Pennebaker and J. L. Mitchell, "JPEG: Still Image Data Compression Standard". Van Nostrand Reinhold, New York, 1993.

[11] M. A. Robertson, "Computationally-Efficient Post-Processing of Compressed Video Streams", Master's Thesis, University of Notre Dame, Notre Dame IN, 1998.

[12] R. R. Schultz and R. L. Stevenson, "Improved definition image expansion", Proceedings 1992 Int. Conf. Acoust., Speech and Signal Process., San Francisco, pp. III:173-176, Mar. 1992.

[13] R. L. Stevenson and S. M. Schweizer, "A nonlinear filtering structure for image smoothing in mixed noise environments", Journal of Mathematical Imaging and Vision, vol. 2, pp. 137-154, Nov. 1992.

3 IMAGE RECOVERY FROM COMPRESSED VIDEO USING MULTICHANNEL REGULARIZATION

Yongyi Yang, Mungi Choi, and Nikolas P. Galatsanos

Department of Electrical and Computer Engineering
Illinois Institute of Technology
Chicago, IL 60616
{yy,mgc,npg}@ece.iit.edu

Abstract: In this chapter we propose a multichannel recovery approach to ameliorate coding artifacts in compressed video. The main shortcomings of previously proposed recovery algorithms for this problem was that only spatial smoothness was explicitly enforced. In this chapter we attempt to ameliorate the above problem. According to the proposed approach, prior knowledge is enforced by introducing regularization operators which complement the transmitted data. The term multichannel implies that both the spatial (within-channel) and temporal (across-channel) properties of the image sequences are used in the recovery process. More specifically, regularization operators are defined that in addition to spatial smoothness *explicitly* enforce smoothness along the motion trajectories. Since the compressed images due to quantization belong to known convex sets, iterative gradient projection algorithms are proposed to minimize the regularized functional and simultaneously guarantee membership to these sets. Numerical experiments are shown using H.261 and H.263 coded sequences. These experiments demonstrate that introduction of temporal regularization offers a significant improvement both visually and from a peak signal-to-noise ratio point of view.

3.1 INTRODUCTION

With the growing demand for efficient digital representation of image and video in a wide range of applications, image and video compression has attracted considerable attention in recent years. Evidence of this is the emergence of several international image and video coding standards such as JPEG [1], MPEG-1 [2], MPEG-2 [3], H.261 [4], H.263 [5], and the ongoing work of MPEG-4 [6]. What makes compression of images and video possible is the well-known fact that there exists a great deal of information redundancy in image and video, and every efficient coding scheme attempts to exploit this redundancy.

The information redundancy in a video sequence is divided into spatial redundancy and temporal redundancy. The spatial redundancy exhibits as the similarity of neighboring pixels in an image, whereas the temporal redundancy exhibits as similarity between neighboring frames in a video sequence. A popular video coding scheme is to use a transform, such as a subband/wavelet transform or discrete cosine transform (DCT), to exploit the spatial redundancy, and use motion compensated prediction to take advantage of the temporal redundancy. For example, the block discrete cosine transform (BDCT) and motion compensation are widely used in the existing video coding standards.

An inherent problem with image and video compression is that it results in various undesirable coding artifacts, especially in low bit-rate applications. For example, coding artifacts such as "blocking", "ringing effects", and "mosquito noise" are well known to exist in JPEG and MPEG compressed images and video. In the literature various postprocessing and recovery algorithms have been proposed in attempt to ameliorate these compression artifacts in the compressed images. For example, in [7, 9, 10, 11, 12] different filtering algorithms are used to reduce the artifacts in a compressed image, while in [13, 15, 16, 17, 18, 19, 20] image recovery approaches are proposed to reconstruct the compressed image so that it is free (or nearly free) of artifacts. The essence of these recovery algorithms is to reconstruct the image by taking advantage of, in addition to the available received compressed data, the existence of strong correlation among the pixels of the image (i.e., spatial redundancy) either through assuming an underlying *a priori* probability model or by imposing a smoothness constraint on the image.

For the recovery of compressed video, a straightforward approach would be to treat each frame in a video as an independent image and recover it separately from the other frames. However, such an approach is only suboptimal since it does not make use of the existence of strong interframe correlation (i.e., temporal redundancy) in a video sequence. It is well known from other image sequence recovery problems that the incorporation of interframe correlations can lead to significant improvements, see for example [14]. Indeed, our initial study in [21] demonstrates that significant improvement in the quality of the recovery images can be achieved when the temporal correlation in a video sequence is utilized in a recovery algorithm. The recovery algorithm in [21] is multichannel in nature in that it enforces, along with spatial smoothness, temporal smoothness (i.e. interframe correlation) along the motion trajectories by utilizing the motion information that is readily available in the compressed data. This algorithm, however, is somewhat limited and primitive in the way that the motion information is utilized. Specifically, it can utilize only unidirectional (forward) motion information and assumes that the motion compensation is done at an integer-pixel level. In this paper we extend this work so that the recovery algorithm can fully utilize the motion information more commonly seen in the currently existing video coding standards such as MPEG and H.263 compression, where bi-directional motion information is available and the motion compensation can be done at a sub-pixel level.

The rest of the chapter is organized as follows: In Section 2, we first introduce some notion on video coding, particularly the coding scheme of block DCT and motion compensation. Then, we give motivation on the formulation of video decoding as an ill-posed recovery problem by using the principle of regularization. The regularization terms which utilize the spatial domain correlation and the temporal domain correlation are defined in Section 3. The recovery algorithm, which is based on the gradient projection algorithm, is derived in Section 4. Also furnished in Section 4 is a numerically simplified version of this recovery algorithm, which is demonstrated later on that it can, nevertheless, achieve comparable performance with the original algorithm. In Section 5, some issues regarding the implementation of the recovery algorithms are discussed and numerical results are presented. Finally, conclusions are given in Section 6.

3.2   BACKGROUND AND MOTIVATION

In this section we give the motivation that decoding of compressed video can be treated naturally as a recovery problem. We first review some notion on video coding. To help remain focused, we'll concentrate on a motion compensated transform coding scheme. Notice that such a coding scheme is widely used in the currently existing coding standards such as MPEG, H.261, and H.263.

3.2.1   Motion compensated video coding

In motion compensated video coding such as MPEG, the frames in a video sequence are broadly classified into intra-coded frames and predictively coded frames. While an intra-coded frame, also known as an anchor frame, is coded using a transform such as block DCT, a predictively coded frame is first predicted (either unidirectionally or bi-directionally) from the most recently reconstructed reference frame(s), and the prediction error is then compressed by using the coding transform. In MPEG coding, for example, the block DCT is used, and an intra-coded frame is known as an I-frame, and a forward (or bidirectional) predictive coded frame is called a P-frame (or B-frame).

Some notation is due in order to quantitatively describe this coding process. Let $\mathbf{f}_1, \mathbf{f}_2, \cdots, \mathbf{f}_L$ denote L frames of a video sequence, where $\mathbf{f}_l$ is a vector representation of frame l through a lexicographical ordering of its pixels. For frame $\mathbf{f}_l$, define

$$\bar{\mathbf{f}}_l \triangleq \begin{cases} \mathbf{f}_l & \text{if } \mathbf{f}_l \text{ is intra coded} \\ \mathbf{f}_l - \hat{\mathbf{f}}_l & \text{if } \mathbf{f}_l \text{ is predictively coded,} \end{cases} \tag{3.1}$$

where $\hat{\mathbf{f}}_l$ denotes the motion compensated prediction of the frame $\mathbf{f}_l$ from the most recently reconstructed reference frame(s). That is, $\bar{\mathbf{f}}_l$ is simply the motion compensated error of the frame $\mathbf{f}_l$ when it is predictively coded. This prediction error $\bar{\mathbf{f}}_l$ is then further compressed by using the coding transform (such as block DCT), denoted by T, followed by the quantizer, denoted by $\mathcal{Q}$, to obtain the compressed data, say $\bar{\mathbf{F}}_l$. In other words, $\bar{\mathbf{F}}_l$ denotes the quantized transform coefficients of $\bar{\mathbf{f}}_l$. In short, we have

$$\bar{\mathbf{F}}_l = Q T_B \bar{\mathbf{f}}_l \tag{3.2}$$

for $l = 1, 2, \cdots, L$. These quantized data, along with the necessary motion information, are then entropy coded and transmitted to the receiver.

3.2.2 Video decoding as video recovery

The task at the receiver is to reconstruct the images $\mathbf{f}_1, \mathbf{f}_2, \cdots, \mathbf{f}_L$ from the received data. Since quantization is typically a many-to-one mapping, i.e., the operator $\mathcal{Q}$ is not invertible, we can no longer determine exactly the original images $\mathbf{f}_1, \mathbf{f}_2, \cdots, \mathbf{f}_L$ from the received quantized coefficients $\bar{\mathbf{F}}_1, \bar{\mathbf{F}}_2, \cdots, \bar{\mathbf{F}}_L$. In a classical decoder, these quantized coefficients are simply taken as the transform coefficients and operations opposite to the coding process are taken to obtain the compressed images. That is, the frames $\mathbf{f}_l, l = 1, 2, \cdots, L$, are decoded as

$$\mathbf{g}_l \triangleq \begin{cases} T^{-1}\bar{\mathbf{F}}_l & \text{if } \mathbf{f}_l \text{ is intra coded} \\ \hat{\mathbf{f}}_l + T^{-1}\bar{\mathbf{F}}_l & \text{if } \mathbf{f}_l \text{ is predictively coded,} \end{cases} \tag{3.3}$$

where T^{-1} denotes the inverse of the coding transform. As pointed out earlier, these decoded images exhibit, especially at low bit-rates, various undesirable artifacts, which is due to the information lossy nature of the quantization process.

The information loss caused by quantization can be quantitatively described. Since the quantized value of each coefficient specifies an interval that the exact value of that coefficient should belong to, the knowledge of the quantized coefficients $\bar{\mathbf{F}}_l$ defines the following set to which the frame $\mathbf{f}_l$ should belong [13, 15]:

$$C_l \triangleq \left\{ \mathbf{f}_l : \alpha_n^{min} \le \left(T\bar{\mathbf{f}}_l\right)_n \le \alpha_n^{max}, n \in \mathcal{I} \right\}, \tag{3.4}$$

where $\left(T\bar{\mathbf{f}}_l\right)_n$ is used to denote the n-th transform coefficient of $\bar{\mathbf{f}}_l$, α_n^{min} and α_n^{max} are the end-points of the quantization interval associated with this coefficient, and $\mathcal{I}$ denotes the inx set of the pixels in the frame.

Clearly, every element in C_l will result in the same quantized data as the original image $\mathbf{f}_l$ does. As a matter of fact, one can quickly verify that the decoded image $\mathbf{g}_l$ in Eq. (3.3) also belongs to C_l, yet it exhibits coding artifacts. Thus, without additional knowledge it is impossible to determine the original image exactly. Therefore, the recovery of the images $\mathbf{f}_l, l = 1, 2, \cdots, L$, from the received data becomes an ill-posed problem.

An effective method for obtaining satisfactory solutions to ill-posed problems is the method of regularization [22]. It has been applied successfully to solve a number of signal and image recovery problems. In particular it is used in [15] for the reduction of blocking artifacts in compressed images. According to this approach an objective function is defined to consist of two terms whereby one term is used to express the fidelity of the solution to the available data and the other is to incorporate prior knowledge or impose desired properties on the solution. The solution to the problem is then obtained through the minimization of this objective function. For the recovery of video from its compressed data using the method of regularization, the key is to define the objective function in such a way that its regularization terms can take advantage of, in addition to the received data, both the spatial domain and the temporal domain correlations in the video. This will be the focus of the next section.

3.3   THE REGULARIZATION FUNCTION

Due to the need to treat the frames $\mathbf{f}_1, \mathbf{f}_2, ..., \mathbf{f}_L$ as a whole in a recovery algorithm, we use in the following the symbol $\mathbf{F}$ to denote the L-tuple of $\mathbf{f}_1, \mathbf{f}_2, ..., \mathbf{f}_L$, i.e.,

$$\mathbf{F} \overset{\triangle}{=} \left(\mathbf{f}_1, \mathbf{f}_2, ..., \mathbf{f}_L\right). \tag{3.5}$$

The existence of strong spatial and temporal correlations in $\mathbf{F}$ suggests the use of an objective function of the following form for the recovery of $\mathbf{F}$ from its compressed data:

$$J(\mathbf{F}) \overset{\triangle}{=} \left(\sum_{l=1}^{L} \|\mathbf{f}_l - \mathbf{g}_l\|^2\right) + \lambda_1 E_s(\mathbf{F}) + \lambda_2 E_t(\mathbf{F}), \tag{3.6}$$

where three regularization terms are involved, each of which is explained below:

- In the first term, $\|\cdot\|$ denotes the Euclidean norm, and $\mathbf{g}_l$ the compressed images as defined in Eq. (3.3). The first term is used to enforce the fidelity of the recovered images $\mathbf{f}_1, \mathbf{f}_2, ..., \mathbf{f}_L$ to the received data, which are represented in the spatial domain by the images $\mathbf{g}_l, l = 1, 2 \cdots, L$.

- The second term $E_s(\mathbf{F})$ and the third term $E_t(\mathbf{F})$, which are to be defined in the following, are used to enforce the spatial and the temporal correlations in $\mathbf{F}$, respectively.

- The constants λ_1 and λ_2 are called regularization parameters and their role is to balance the influence of their associated regularization terms on the objective function.

The strong spatial domain correlation in the images $\mathbf{F}$ is indicated by the fact that neighboring pixels in a frame are typically somewhat similar, and the image as a result is often perceived as being *smooth*. The term $E_s(\mathbf{F})$ in the objective function $J(\mathbf{F})$ in Eq. (3.6) is used to exploit this property to enforce the spatial correlation in the recovered images. Specifically, it is defined in such a way that it will penalize any deviation from smoothness by the recovered images $\mathbf{F}$. Note that the spatial smoothness property of images has been exploited in almost every proposed processing algorithm, ranging from simple filtering to the more sophisticated probability model based recovery approaches, for the reduction of compression artifacts. In [15, 16, 20], for example, smoothness constraints of different forms are directly imposed on the image in the recovery algorithms to effectively remove various artifacts in compressed images.

The temporal domain correlation, on the other hand, is signified by the similarity between the neighboring frames in $\mathbf{F}$. The term $E_t(\mathbf{F})$ in Eq. (3.6) is defined in the following to exploit this property by enforcing smoothness along the motion trajectories in the frames $\mathbf{F}$. That is, $E_t(\mathbf{F})$ is to enforce *temporal smoothness* between the frames, in contrast to the spatial smoothness as enforced by $E_s(\mathbf{F})$.

3.3.1 Spatial domain regularization

In block transform based video coding such as in MPEG, images are typically coded on a block by block basis, where a particular block is processed independent of the others. As a result, discontinuities will occur at coding block boundaries. This discontinuity exhibits as annoying blocking artifacts at low bit rates. Indeed, it is observed in [15] that the local variation of the pixels at the coding block boundaries of an compressed image tends to be significantly larger than that of the pixels inside the coding blocks. Thus, special attention is called for at the coding block boundaries in the definition of $E_s(\mathbf{F})$ in order to effectively suppress the blocking artifacts. With this in mind, we in the following define $E_s(\mathbf{F})$ to include two terms: one enforces smoothness at the coding boundaries and the other enforces smoothness inside the coding blocks.

First, let's introduce some operators to separate the spatial variation of an image at its coding block boundaries from that inside its coding blocks. We will assume that $\mathbf{f}$ is an $M \times N$ image with pixels $f(i, j), i = 1, 2, \cdots, M, j = 1, 2, \cdots, N$, and that a coding block size of 8×8 is used. Note that the frame inx l has been submerged in $\mathbf{f}$ for the sake of notational simplicity.

A. Between-block variation   The total variation at the coding block boundaries of an image is further decomposed into two terms: the variation at the *vertical* block-boundaries and that at the *horizontal* block-boundaries. For the former, define Q_{VB} to be such an operator that it finds the difference between adjacent columns at the block boundaries of an image. That is,

$$(Q_{VB}\mathbf{f})(i,j) \triangleq \begin{cases} f(i,j) - f(i,j+1) & \text{if } j = 8l \text{ for some } l = 1, 2, \cdots, N/8 - 1 \\ 0 & \text{otherwise,} \end{cases}$$

(3.7)

where $(Q_{VB}\mathbf{f})(i,j)$ denotes the value of $Q_{VB}\mathbf{f}$ at (i,j). Then, we have

$$\|Q_{VB}\mathbf{f}\|^2 = \sum_{i=1}^{M} \sum_{l=1}^{N/8-1} [f(i,8l) - f(i,8l+1)]^2 .$$

(3.8)

Clearly, the quantity $\|Q_B\mathbf{f}\|^2$ captures the total variation between the adjacent columns at the block boundaries of the image $\mathbf{f}$.

In a similar fashion, we can define an operator Q_{HB} to find the difference at the *horizontal* coding block boundaries of an image. Then, the quantity $\|Q_{VB}\mathbf{f}\|^2 + \|Q_{HB}\mathbf{f}\|^2$ as a whole is a measure of discontinuity (hence, the blocking artifacts) at the coding block boundaries of an image $\mathbf{f}$. Note that the operators Q_{VB} and Q_{HB} are used in [15] to define smoothness constraint sets to suppress blocking artifact of an image in the vertical and horizontal directions, respectively.

B. Within-block variation   Like that at the coding block boundaries of an image, the total variation inside the coding blocks is also decomposed into two terms: the variation between the adjacent vertical columns and that between the adjacent horizontal rows. For the former, let $Q_{V\bar{B}}$ to be such an operator that

$$(Q_{V\bar{B}}\mathbf{f})(i,j) \triangleq \begin{cases} f(i,j) - f(i,j+1) & \text{if } j \neq 8l \text{ for any } l = 1, 2, \cdots, N/8 - 1 \\ 0 & \text{otherwise.} \end{cases}$$

(3.9)

In words, $Q_{V\bar{B}}$ finds the differences between all the adjacent columns inside the coding blocks of the image $\mathbf{f}$.

Similarly, we can define an operator, say $Q_{H\bar{B}}$, to find the differences between all the adjacent rows inside the coding blocks of an image. Then, the quantity $\|Q_{V\bar{B}}\mathbf{f}\|^2 + \|Q_{H\bar{B}}\mathbf{f}\|^2$ as a whole gives the total variation inside the coding blocks of the image $\mathbf{f}$.

C. The spatial regularization term $E_s(\mathbf{F})$ With the spatial variation operators introduced above, the spatial regularization term $E_s(\mathbf{F})$ on the images $\mathbf{F} = (\mathbf{f}_1, \mathbf{f}_2, ..., \mathbf{f}_L)$

is finally defined as

$$E_s(\mathbf{F}) \triangleq \sum_{l=1}^{L} \left[\left(\|Q_{VB}\mathbf{f}_l\|^2 + \|Q_{HB}\mathbf{f}_l\|^2 \right) + \beta \left(\|Q_{V\bar{B}}\mathbf{f}_l\|^2 + \|Q_{H\bar{B}}\mathbf{f}_l\|^2 \right) \right], \quad (3.10)$$

where the first part penalizes the variations at the coding block boundaries of the frames $\mathbf{f}_1, \mathbf{f}_2, ..., \mathbf{f}_L$, and the second part penalizes the variations inside their coding blocks. The parameter β in Eq. (3.10) is a constant chosen to emphasize the differential treatment of the smoothness at the coding block boundaries from that inside the coding blocks so as to effectively remove blocking artifacts in the recovered images $\mathbf{f}_1, \mathbf{f}_2, ..., \mathbf{f}_L$.

Observe that each term involved in Eq. (3.10) is a quadratic and convex function of the image vector $\mathbf{f}_l$. As a result, the regularization term $E_s(\mathbf{F})$ is a quadratic and convex function of $\mathbf{F} = (\mathbf{f}_1, \mathbf{f}_2, ..., \mathbf{f}_L)$.

3.3.2 Temporal domain regularization

The definition of the spatial domain regularization term $E_s(\mathbf{F})$ above essentially captures the differences between *spatially* neighboring pixels. By the same token, we can define the temporal domain regularization term $E_t(\mathbf{F})$ based on the differences between *temporally* neighboring frames. Unfortunately, the direct frame-to-frame difference cannot be used because of the relative motion between the pixels in neighboring frames. Instead, the motion compensated frame difference (MFD) should be used. For two neighboring frames $\mathbf{f}_{l-1}$ and $\mathbf{f}_l$, their MFD is defined as

$$MFD(\mathbf{f}_{l-1}, \mathbf{f}_l) \triangleq \sum_{i=1}^{M} \sum_{j=1}^{N} \left[f_l(i,j) - f_{l-1}(i+m^{(i,j)}, j+n^{(i,j)}) \right]^2, \quad (3.11)$$

where $(m^{(i,j)}, n^{(i,j)})$ denotes the displacement vector of the pixel (i,j) of frame $\mathbf{f}_l$ relative to its correspondence in frame $\mathbf{f}_{l-1}$. In other words, $(m^{(i,j)}, n^{(i,j)})$ is the motion vector of the pixel (i,j) in frame $\mathbf{f}_l$ with respect to frame $\mathbf{f}_{l-1}$.

Note that the motion vectors $(m^{(i,j)}, n^{(i,j)})$ may not be of integer precision, and as a result the corresponding $f_{l-1}(i+m^{(i,j)}, j+n^{(i,j)})$ in Eq. (3.11) may not point to an exact pixel location in the reference frame $\mathbf{f}_{l-1}$. For example, half-pixel accuracy motion vectors are allowed in MPEG. In such a case, $f_{l-1}(i + m^{(i,j)}, j + n^{(i,j)})$ needs to be interpolated from its closest neighboring pixels. As an example, for $(m^{(i,j)}, n^{(i,j)}) = (1.5, -2.5)$,

$$\begin{aligned} f_{l-1}(i+1.5, j-2.5) &= \frac{1}{4} \left[f_{l-1}(i+1, j-2) + f_{l-1}(i+1, j-2) \right. \\ &\quad + \left. f_{l-1}(i+2, j-3) + f_{l-1}(i+2, j-3) \right] \end{aligned} \quad (3.12)$$

From its definition in Eq. (3.11), it is clear that $MFD(\mathbf{f}_{l-1}, \mathbf{f}_l)$ captures the total variation of the pixels in frame $\mathbf{f}_l$ from their counterparts in frame $\mathbf{f}_{l-1}$ along their motion trajectories. For the purpose of enforcing temporal correlation, the regularization term $E_t(\mathbf{f})$ can then be defined as

$$E_t(\mathbf{F}) \triangleq \sum_{l=2}^{L} MFD(\mathbf{f}_{l-1}, \mathbf{f}_l). \tag{3.13}$$

In the definition of $DFD(\mathbf{f}_{l-1}, \mathbf{f}_l)$ the knowledge of the motion vectors $(m^{(i,j)}, n^{(i,j)})$ is required. These motion vectors are typically available in the compressed data when the frames are forward predictively coded, i.e., the frame $\mathbf{f}_l$ is predicted using the frame $\mathbf{f}_{l-1}$. In practice, some pixels in $\mathbf{f}_l$ may not be predictable from those in $\mathbf{f}_{l-1}$, for example, when they belong to a newly uncovered region in $\mathbf{f}_l$ due to motion. In MPEG coding, for example, some macroblocks in a predictively coded frame may actually be intra-coded, i.e., coded without motion compensated. In such a case, they should accordingly be eliminated from the definition of MFD in Eq. (3.11). That is, the motion compensated frame difference should now be modified as

$$MFD(\mathbf{f}_{l-1}, \mathbf{f}_l) \triangleq \sum_{(i,j) \in \mathcal{I}_p} \left[f_l(i,j) - f_{l-1}(i+m^{(i,j)}, j+n^{(i,j)}) \right]^2, \tag{3.14}$$

where $\mathcal{I}_p$ is the inx set of pixels in $\mathbf{f}_l$ that are predictively coded from $\mathbf{f}_{l-1}$.

The regularization term $E_t(\mathbf{F})$ in Eq. (3.13) is readily applicable to coding schemes such as H.261 and H.263 where only forward predictive coding is used. In a coding scheme such as MPEG where both forward predictive coding and bi-directional predictive coding are allowed, $E_t(\mathbf{f})$ needs to be modified accordingly in order to fully take advantage of the available motion information. Specifically, for the case of forward predictive coding in MPEG, say, frame $\mathbf{f}_l$ is forward predicted from reference frame $\mathbf{f}_{l-2}$, the corresponding MFD term in $E_t(\mathbf{F})$ should now be adjusted to $MFD(\mathbf{f}_{l-2}, \mathbf{f}_l)$. The case of bi-directional predictive coding in MPEG is somewhat more involved. Suppose, for example, that frame $\mathbf{f}_l$ is bi-directionally predicted from reference frames $\mathbf{f}_{l-2}$ and $\mathbf{f}_{l+1}$. Then four possibilities exist in how a particular macroblock in $\mathbf{f}_l$ is coded: a) it is intra-coded, i.e., coded without prediction; b) it is forward predicted from frame $\mathbf{f}_{l-2}$; c) it is backward predicted from frame $\mathbf{f}_{l+1}$; and finally d) it is bi-directionally predicted from both $\mathbf{f}_{l-2}$ and $\mathbf{f}_{l+1}$. In such a case, we should include the following two MFD terms in $E_t(\mathbf{F})$ to utilize the motion information in $\mathbf{f}_l$:

$$MFD(\mathbf{f}_{l-2}, \mathbf{f}_l) + MFD(\mathbf{f}_{l+1}, \mathbf{f}_l) \tag{3.15}$$

where the first term is defined for the macroblocks in $\mathbf{f}_l$ that are either forward or bi-directionally predicted, while the second term is for the macroblocks that are either backward or bi-directionally predicted. Clearly, for a macroblock in $\mathbf{f}_l$ that is bi-directionally predicted, it will appear in both terms in Eq. (3.15), and hence both its forward and backward motion information is utilized in Eq. (3.15).

In the following we use two examples to illustrate the exact forming of the temporal regularization term $E_t(\mathbf{F})$ when different modes of predictive coding are used.

Example 1: Assume that a video is compressed using forward predictive coding (such as H.261 or H.263), and its first a few frames are coded in the following pattern:

$$1(I) \quad 2(P) \quad 3(P) \quad 4(P) \quad 5(P) \quad 6(P) \quad 7(P) \quad 8(P) \quad \cdots \qquad (3.16)$$

That is, frame #1 is intra-coded, frame #2 is predictively coded from frame #1, frame #3 is predictively coded from frame #2, and so forth. Suppose that $L = 7$. In such a case, $E_t(\mathbf{F})$ in Eq. (3.13) can be directly used, i.e.,

$$E_t(\mathbf{F}) = MFD(\mathbf{f}_1, \mathbf{f}_2) + MFD(\mathbf{f}_2, \mathbf{f}_3) + \cdots + MFD(\mathbf{f}_5, \mathbf{f}_6) + MFD(\mathbf{f}_6, \mathbf{f}_7). \quad (3.17)$$

Example 2: Assume that a video is compressed using both forward and bi-directional predictive coding (such as MPEG), and its first a few frames are coded in the following pattern:

$$1(I) \quad 2(B) \quad 3(B) \quad 4(P) \quad 5(B) \quad 6(B) \quad 7(P) \quad 8(B) \quad \cdots \qquad (3.18)$$

In words, described in their encoding order, frame #1 is intra-coded; frame #4 is predictively coded from frame #1; frame #2 is then bi-directionally coded from frames #1 and #4, and so is frame #3; frame #7 then is predictively coded from frame #4; frame #5 is then bi-directionally coded from frames #4 and #7, and so forth. In such a case, the definition of $E_t(\mathbf{F})$ in Eq. (3.13) is no longer directly applicable. Instead, the MFD terms in Eq. (3.15) should be used for the B-pictures. Again, assume that $L = 7$ is used. Then, we have

$$
\begin{aligned}
E_t(\mathbf{F}) \quad = \quad & \underbrace{MFD(\mathbf{f}_1, \mathbf{f}_4)}_{\text{frame 4}} + \underbrace{MFD(\mathbf{f}_1, \mathbf{f}_2) + MFD(\mathbf{f}_4, \mathbf{f}_2)}_{\text{frame 2}} \\
& + \underbrace{MFD(\mathbf{f}_1, \mathbf{f}_3) + MFD(\mathbf{f}_4, \mathbf{f}_3)}_{\text{frame 3}} + \underbrace{MFD(\mathbf{f}_4, \mathbf{f}_7)}_{\text{frame 7}} \\
& + \underbrace{MFD(\mathbf{f}_4, \mathbf{f}_5) + MFD(\mathbf{f}_7, \mathbf{f}_5)}_{\text{frame 5}} \\
& + \underbrace{MFD(\mathbf{f}_4, \mathbf{f}_6) + MFD(\mathbf{f}_7, \mathbf{f}_6)}_{\text{frame 6}}. \qquad (3.19)
\end{aligned}
$$

Note that in Eq. (3.19), the MFD terms associated with each predicted frame are clearly indicated.

Finally, we point out that due to the quadratic and convex nature of the MFD terms used in its definition, the regularization term $E_t(\mathbf{F})$, be it of various forms, is also quadratic and convex in terms of $\mathbf{F} = (\mathbf{f}_1, \mathbf{f}_2, ..., \mathbf{f}_L)$.

Another form of the temporal regularization term that has been shown to be very effective in image restoration problems [14] is

$$E_t(\mathbf{F}) = \sum_{i=1}^{l} ||\mathbf{f}_i - \hat{\mathbf{f}}_i^{MC}||^2, \qquad (3.20)$$

where $\hat{\mathbf{f}}_i^{MC}$ is the motion compensated estimate of frame $\mathbf{f}_i$. The main difference between the temporal regularization term in Eq. (3.13) and (3.20) is that for the latter in cases that more than one frame are used for the prediction of $\mathbf{f}_i$ the cross-correlation of the motion compensated errors are also forced to be small. More specifically, assume that frame $\mathbf{f}_l$ is bi-directionally predicted from frames $\mathbf{f}_k$ and $\mathbf{f}_{k'}$. Then, we have

$$||\mathbf{f}_l - \hat{\mathbf{f}}_l^{MC}||^2 = ||\mathbf{f}_l - \tfrac{1}{2}(\hat{\mathbf{f}}_k^l + \hat{\mathbf{f}}_{k'}^l)||^2 = \quad \tfrac{1}{4}MFD(\mathbf{f}_k, \mathbf{f}_l) + \tfrac{1}{4}MFD(\mathbf{f}_{k'}, \mathbf{f}_l) + \\ \tfrac{1}{2}(\mathbf{f}_l - \hat{\mathbf{f}}_k^l)^T(\mathbf{f}_l - \hat{\mathbf{f}}_{k'}^l)$$

$$(3.21)$$

where $\hat{\mathbf{f}}_k^l$ and $\hat{\mathbf{f}}_{k'}^l$ are the motion compensated estimates of frame l using frames k and k', respectively. The additional term $\tfrac{1}{2}(\mathbf{f}_l - \hat{\mathbf{f}}_k^l)^T(\mathbf{f}_l - \hat{\mathbf{f}}_{k'}^l)$ that appears corresponds to the previously mentioned cross-correlation.

3.4 THE RECOVERY ALGORITHMS

Now that the regularized objective function $J(\mathbf{F})$ in Eq. (3.6) is defined, the recovery of video $\mathbf{F} = (\mathbf{f}_1, \mathbf{f}_2, ..., \mathbf{f}_L)$ from its compressed data can then be solved by directly minimizing this objective function. Recall from our earlier discussion in Section 2, however, that each original frame $\mathbf{f}_l$ is known to belong to a constraint set C_l defined by the received data in Eq. (3.4). It is therefore more reasonable to seek the solution from the following constrained problem:

$$\min \ J(\mathbf{F}) = \sum_{l=1}^{L} ||\mathbf{f}_l - \mathbf{g}_l||^2 + \lambda_1 E_s(\mathbf{F}) + \lambda_2 E_t(\mathbf{F})$$

$$\text{subject to } \mathbf{f}_l \in C_l, l = 1, 2 \cdots, L. \qquad (3.22)$$

Note that the constraint sets C_l on the individual frames $\mathbf{f}_l, l = 1, 2, \cdots, L$, can be written equivalently as a single constraint on the whole sequence $\mathbf{F} = (\mathbf{f}_1, \mathbf{f}_2, ..., \mathbf{f}_L)$. Specifically, define

$$C \overset{\triangle}{=} \{(\mathbf{f}_1, \mathbf{f}_2, ..., \mathbf{f}_L) : \ \mathbf{f}_l \in C_l, l = 1, 2 \cdots, L\}. \qquad (3.23)$$

Then, the statement $\mathbf{f}_l \in C_l, l = 1, 2 \cdots, L$, is simply equivalent to $\mathbf{F} \in C$. Thus, the optimization problem in Eq. (3.22) is equivalently written as

$$\min J(\mathbf{F}) \quad \text{subject to} \ \ \mathbf{F} \in C. \qquad (3.24)$$

Recall that earlier in the previous section both $E_s(\mathbf{F})$ and $E_t(\mathbf{F})$ were pointed out to be quadratic and convex functions of $\mathbf{F}$. As a result, the objective function $J(\mathbf{F})$ is also quadratic and convex in terms of $\mathbf{F}$. Also, the sets $C_l, l = 1, 2 \cdots, L$, are closed and convex [15, 20], and consequently so is the set C. Therefore, the optimization problem in Eq. (3.24) is in the form of a convex functional under a convex constraint. It is well known that a problem of such a form has a well-defined unique solution, and this solution can be found using a so-called iterative gradient projection algorithm [23], i.e.,

$$\mathbf{F}^{k+1} = P_C \left(\mathbf{F}^k - \alpha \nabla J(\mathbf{F}^k) \right), \quad k = 0, 1, 2, \cdots \tag{3.25}$$

where P_C is the projection operator onto the set C, and α a relaxation parameter that controls the rate of convergence of the iteration. Also in Eq. (3.25), $\mathbf{F}^k$ denotes the estimate of $\mathbf{F}$ after the k-th iteration, and $\nabla J(\mathbf{F}^k)$ is the gradient of $J(\mathbf{F})$ with respect to $\mathbf{F}$ evaluated at $\mathbf{F}^k$. The computation of the projector P_C and the gradient $\nabla J(\mathbf{F}^k)$ is explained in detail in the following.

3.4.1 The projection operator P_C

From its definition in Eq. (3.23), it is easy to see that for $\mathbf{F} = (\mathbf{f}_1, \mathbf{f}_2, ..., \mathbf{f}_L)$ its projection $P_C \mathbf{F}$ onto the set C is simply given by the projections of the individual $\mathbf{f}_l$ onto their corresponding sets C_l. That is,

$$P_C \mathbf{F} = (P_{C_1} \mathbf{f}_1, P_{C_2} \mathbf{f}_2, \cdots, P_{C_L} \mathbf{f}_L). \tag{3.26}$$

The individual projection of $\mathbf{f}_l$ onto the set C_l is well-known [15, 20]. For T an orthonormal coding transform such as the block DCT, the projection $P_{C_l} \mathbf{f}_l$ is given by

$$P_{C_l} \mathbf{f}_l = \begin{cases} T^{-1} \bar{\mathbf{F}}_l & \text{if } \mathbf{f}_l \text{ is intra-coded} \\ \bar{\mathbf{f}}_l + T^{-1} \bar{\mathbf{F}}_l & \text{if } \mathbf{f}_l \text{ is predictively coded,} \end{cases} \tag{3.27}$$

where the n-th coefficient of $\bar{\mathbf{F}}_l$ is given by

$$(\bar{\mathbf{F}}_l)_n = \begin{cases} \alpha_n^{min} & \text{if } (T\bar{\mathbf{f}}_l)_n < \alpha_n^{min} \\ \alpha_n^{max} & \text{if } (T\bar{\mathbf{f}}_l)_n > \alpha_n^{max} \\ (T\bar{\mathbf{f}}_l)_n & \text{otherwise.} \end{cases} \tag{3.28}$$

3.4.2 The gradient $\nabla J(\mathbf{F})$

The gradient $\nabla J(\mathbf{F})$ can be expressed in terms of the partial derivatives of $J(\mathbf{F})$ with respect to the individual frames $\mathbf{f}_l$ as

$$\nabla J(\mathbf{F}) = \left(\frac{\partial J(\mathbf{F})}{\partial \mathbf{f}_1}, \frac{\partial J(\mathbf{F})}{\partial \mathbf{f}_2}, \cdots, \frac{\partial J(\mathbf{F})}{\partial \mathbf{f}_L} \right). \tag{3.29}$$

Switching the frame inx from l to k in $J(\mathbf{F})$ in Eq. (3.6), we have

$$J(\mathbf{F}) = \sum_{k=1}^{L} \|\mathbf{f}_k - \mathbf{g}_k\|^2 + \lambda_1 E_s(\mathbf{F}) + \lambda_2 E_t(\mathbf{F}). \tag{3.30}$$

For a particular frame $\mathbf{f}_l, l = 1, 2, \cdots, L$,

$$\frac{\partial J(\mathbf{F})}{\partial \mathbf{f}_l} = 2(\mathbf{f}_l - \mathbf{g}_l) + \lambda_1 \frac{\partial}{\partial \mathbf{f}_l} E_s(\mathbf{F}) + \lambda_2 \frac{\partial}{\partial \mathbf{f}_l} E_t(\mathbf{F}). \tag{3.31}$$

In the following, we consider the computation of the partial derivative terms $\frac{\partial}{\partial \mathbf{f}_l} E_s(\mathbf{F})$ and $\frac{\partial}{\partial \mathbf{f}_l} E_t(\mathbf{F})$ in Eq. (3.31) separately.

A. The partial derivative $\frac{\partial}{\partial \mathbf{f}_l} E_s(\mathbf{F})$ Switching the frame inx from l to k in $E_s(\mathbf{F})$ given in Eq. (3.10), we obtain

$$E_s(\mathbf{F}) = \sum_{k=1}^{L} \left[(\|Q_{VB}\mathbf{f}_k\|^2 + \|Q_{HB}\mathbf{f}_k\|^2) + \beta \left(\|Q_{V\bar{B}}\mathbf{f}_k\|^2 + \|Q_{H\bar{B}}\mathbf{f}_k\|^2 \right) \right].$$
$$\tag{3.32}$$

Then, we have

$$\frac{\partial E_s(\mathbf{F})}{\partial \mathbf{f}_l} = \frac{\partial}{\partial \mathbf{f}_l} \|Q_{VB}\mathbf{f}_l\|^2 + \frac{\partial}{\partial \mathbf{f}_l} \|Q_{HB}\mathbf{f}_l\|^2 + \beta \left(\frac{\partial}{\partial \mathbf{f}_l} \|Q_{V\bar{B}}\mathbf{f}_l\|^2 + \frac{\partial}{\partial \mathbf{f}_l} \|Q_{H\bar{B}}\mathbf{f}_l\|^2 \right).$$
$$\tag{3.33}$$

The partial derivative terms in Eq. (3.33) can be easily computed directly from the definitions of their associated operators. Take the first term for example. From the definition of Q_{VB} in Eq. (3.7), we have

$$\frac{\partial}{\partial f_l(i,j)} \|Q_{VB}\mathbf{f}_l\|^2 =$$

$$\begin{cases} 2[f_l(i,j) - f_l(i,j+1)] & \text{if } j = 8l \text{ for some } l = 1, 2, \cdots, N/8 - 1 \\ 2[f_l(i,j) - f_l(i,j-1)] & \text{if } j = 8l + 1 \text{ for some } l = 1, 2, \cdots, N/8 - 1 \\ 0 & \text{otherwise.} \end{cases} \tag{3.34}$$

Similarly, the other derivative terms in Eq. (3.33) can be derived. The details are omitted for brevity.

B. The partial derivative $\frac{\partial E_t(\mathbf{F})}{\partial \mathbf{f}_l}$ For a particular frame $\mathbf{f}_l$, the partial derivative $\frac{\partial E_t(\mathbf{F})}{\partial \mathbf{f}_l}$ clearly depends on only the MFD terms in $E_t(\mathbf{F})$ that are related to $\mathbf{f}_l$. In general, we have

$$\frac{\partial E_t(\mathbf{F})}{\partial \mathbf{f}_l} = \sum_{k} \frac{\partial}{\partial \mathbf{f}_l} MFD(\mathbf{f}_k, \mathbf{f}_l) + \sum_{k'} \frac{\partial}{\partial \mathbf{f}_l} MFD(\mathbf{f}_l, \mathbf{f}_{k'}), \tag{3.35}$$

where the first summation is over the frames $\mathbf{f}_k$ that are used to predict $\mathbf{f}_l$, and the second is over the frames $\mathbf{f}_{k'}$ that are predicted from $\mathbf{f}_l$. For example, consider $E_t(\mathbf{F})$ defined in Eq. (3.17),

$$E_t(\mathbf{F}) = MFD(\mathbf{f}_1, \mathbf{f}_2) + MFD(\mathbf{f}_2, \mathbf{f}_3) + \cdots + MFD(\mathbf{f}_5, \mathbf{f}_6) + MFD(\mathbf{f}_6, \mathbf{f}_7). \quad (3.36)$$

For $l = 4$, we have

$$\frac{\partial E_t(\mathbf{F})}{\partial \mathbf{f}_4} = \frac{\partial}{\partial \mathbf{f}_4} MFD(\mathbf{f}_3, \mathbf{f}_4) + \frac{\partial}{\partial \mathbf{f}_4} MFD(\mathbf{f}_4, \mathbf{f}_5). \quad (3.37)$$

As another example, consider $E_t(\mathbf{F})$ defined in Eq. (3.19),

$$
\begin{aligned}
E_t(\mathbf{F}) \;=\; & MFD(\mathbf{f}_1, \mathbf{f}_4) + MFD(\mathbf{f}_1, \mathbf{f}_2) + MFD(\mathbf{f}_4, \mathbf{f}_2) \\
& + MFD(\mathbf{f}_1, \mathbf{f}_3) + MFD(\mathbf{f}_4, \mathbf{f}_3) + MFD(\mathbf{f}_4, \mathbf{f}_7) \\
& + MFD(\mathbf{f}_4, \mathbf{f}_5) + MFD(\mathbf{f}_7, \mathbf{f}_5) + MFD(\mathbf{f}_4, \mathbf{f}_6) \\
& + MFD(\mathbf{f}_7, \mathbf{f}_6).
\end{aligned}
\quad (3.38)
$$

For $l = 4$, we have

$$
\begin{aligned}
\frac{\partial E_t(\mathbf{F})}{\partial \mathbf{f}_4} \;=\; & \frac{\partial}{\partial \mathbf{f}_4} MFD(\mathbf{f}_1, \mathbf{f}_4) + \frac{\partial}{\partial \mathbf{f}_4} MFD(\mathbf{f}_4, \mathbf{f}_2) \\
& + \frac{\partial}{\partial \mathbf{f}_4} MFD(\mathbf{f}_4, \mathbf{f}_3) + \frac{\partial}{\partial \mathbf{f}_4} MFD(\mathbf{f}_4, \mathbf{f}_7) \\
& + \frac{\partial}{\partial \mathbf{f}_4} MFD(\mathbf{f}_4, \mathbf{f}_5) + \frac{\partial}{\partial \mathbf{f}_4} MFD(\mathbf{f}_4, \mathbf{f}_6).
\end{aligned}
\quad (3.39)
$$

The evaluation of the partial derivative terms $\frac{\partial}{\partial \mathbf{f}_l} MFD(\mathbf{f}_k, \mathbf{f}_l)$ in Eq. (3.35) is fairly straightforward. Specifically, from the definition of the MFD in Eq. (3.14), we obtain

$$
\frac{\partial}{\partial f_l(i,j)} MFD(\mathbf{f}_k, \mathbf{f}_l) = \begin{cases} 2\left[f_l(i,j) - f_k(i+m^{(i,j)}, j+n^{(i,j)})\right] & \text{if } f_l(i,j) \text{ is predicted} \\ 0 & \text{otherwise,} \end{cases}
\quad (3.40)
$$

where $(m^{(i,j)}, n^{(i,j)})$ is the motion vector of pixel $f_l(i,j)$ with respect to frame $\mathbf{f}_k$.

The evaluation of the partial derivative terms $\frac{\partial}{\partial \mathbf{f}_l} MFD(\mathbf{f}_l, \mathbf{f}_{k'})$ in Eq. (3.35) is somewhat involved. Based on the definition of MFD, we obtain

$$
\frac{\partial}{\partial f_l(i,j)} MFD(\mathbf{f}_l, \mathbf{f}_{k'}) = \sum_{(i',j')} 2\left[f_{k'}(i',j') - \hat{f}_{k'}(i',j')\right] \cdot \frac{\partial \hat{f}_{k'}(i',j')}{\partial f_l(i,j)} \cdot (-1),
\quad (3.41)
$$

where the summation is over those pixels $f_{k'}(i',j')$ in frame $\mathbf{f}_{k'}$ that are predicted from $f_l(i,j)$, and $\hat{f}_{k'}(i',j')$ is simply a short notation for the motion compensated prediction of $f_{k'}(i', j')$. For example, suppose that the pixel $f_l(2, 3)$ has been used to predict pixels $f_{k'}(3, 3)$ and $f_{k'}(3, 4)$ in frame $\mathbf{f}_{k'}$ with

$$\hat{f}_{k'}(3, 3) = \frac{1}{2}\left[f_l(2, 3) + f_l(2, 4)\right], \tag{3.42}$$

and

$$\hat{f}_{k'}(3, 4) = \frac{1}{4}\left[f_l(2, 3) + f_l(2, 4) + f_l(3, 3) + f_l(3, 4)\right]. \tag{3.43}$$

Then,

$$\frac{\partial}{\partial f_l(2, 3)} MFD(\mathbf{f}_l, \mathbf{f}_{k'}) = 2\left[f_{k'}(3, 3) - \hat{f}_{k'}(3, 3)\right]\left(-\frac{1}{2}\right)$$
$$+ 2\left[f_{k'}(3, 4) - \hat{f}_{k'}(3, 4)\right]\left(-\frac{1}{4}\right). \tag{3.44}$$

Clearly, both the computations in Eq. (3.40) and in Eq. (3.41) are quite simple. The only complication in Eq. (3.41), however, is the bookkeeping needed for pixel $f_l(i, j)$ to keep track of the pixels $f_{k'}(i', j')$ that it is used to predict.

Similar observations can be made for the computation of the partial derivatives of $E_t(\mathbf{F})$ when the definition in Eq. (3.20) is used. More specifically, the only difference from the case discussed previously is in finding the derivative of the cross-correlation term. Two cases can be considered. Fist, the derivative $\frac{\partial}{\partial f_k(i,j)}(\mathbf{f}_k - \hat{\mathbf{f}}_l^k)^T(\mathbf{f}_k - \hat{\mathbf{f}}_{l'}^k)$ is straightforward. Taking the derivative of the cross correlation error gives

$$\frac{\partial}{\partial f_k(i, j)}(\mathbf{f}_k - \hat{\mathbf{f}}_l^k)^T(\mathbf{f}_k - \hat{\mathbf{f}}_{l'}^k) =$$

$$\begin{cases} 2f_k(i, j) - [f_l(i + m^{(i,j)}, j + n^{(i,j)}) + f_{l'}(i + p^{(i,j)}, j + q^{(i,j)})] \\ \qquad\qquad \text{if } f_l(i, j) \text{ is predicted} \\ 0 \qquad\qquad\qquad \text{otherwise,} \end{cases} \tag{3.45}$$

where $(m^{(i,j)}, n^{(i,j)})$ and $(p^{(i,j)}, q^{(i,j)})$ are the motion vectors of pixel $f_k(i, j)$ with respect to frames $\mathbf{f}_l$ and $\mathbf{f}_{l'}$ respectively.

The evaluation of the derivative $\frac{\partial}{\partial f_l(i,j)}(\mathbf{f}_k - \hat{\mathbf{f}}_l^k)^T(\mathbf{f}_k - \hat{\mathbf{f}}_{l'}^k)$ is more involved. Based on the definition of the motion compensated estimates of frame $\mathbf{f}_k$ from frames l and l' $\hat{\mathbf{f}}_l^k$ and $\hat{\mathbf{f}}_{l'}^k$, respectively we get

$$\frac{\partial}{\partial f_l(i, j)}(\mathbf{f}_k - \hat{\mathbf{f}}_l^k)^T(\mathbf{f}_k - \hat{\mathbf{f}}_{l'}^k) = \sum_{i', j'}[f_k(i', j') - \hat{f}_{l'}^k(i', j')]\frac{\partial \hat{f}_l^k(i', j')}{\partial f_l(i, j)}(-1) \tag{3.46}$$

where the summation is over all the pixel of $f_k(i', j')$ in frame $\mathbf{f}_k$ that are predicted using $f_l(i, j)$ and $\hat{f}_l^k(i', j')$ is the motion compensated prediction of $f_k(i', j')$ from frame $\mathbf{f}_l$.

3.4.3 The recovery algorithms

Since both the projection P_C in Eq. (3.26) and the gradient $\nabla J(\mathbf{F})$ in Eq. (3.29) can be realized on a frame by frame basis, the recovery algorithm in Eq. (3.25) can be equivalently written in terms of individual frames as

$$\mathbf{f}_l^{k+1} = P_{C_l}\left(\mathbf{f}_l^k - \alpha \frac{\partial J(\mathbf{F}^k)}{\partial \mathbf{f}_l}\right) \tag{3.47}$$

for $l = 1, 2, \cdots, L$. Combining this with Eq. (3.31), we obtain

$$\mathbf{f}_l^{k+1} = P_{C_l}\left[\mathbf{f}_l^k - \alpha\left(2(\mathbf{f}_l - \mathbf{g}_l) + \lambda_1 \frac{\partial}{\partial \mathbf{f}_l} E_s(\mathbf{F}^k) + \lambda_2 \frac{\partial}{\partial \mathbf{f}_l} E_t(\mathbf{F}^k)\right)\right], \tag{3.48}$$

for $l = 1, 2, \cdots, L$. In Eq. (3.48), the term $\frac{\partial}{\partial \mathbf{f}_l} E_s(\mathbf{F}^k)$ is computed according to Eq. (3.33), and the term $\frac{\partial}{\partial \mathbf{f}_l} E_t(\mathbf{F}^k)$ is computed according to Eq. (3.35). The projection operation P_{C_l} in Eq. (3.48) is given in Eq. (3.27) and involves the coding transform. In practice, the coding transform is often implemented with a fast algorithm such as the fast DCT for block DCT. Clearly, all the computations involved in the recovery algorithm in Eq. (3.48) are fairly simple and straightforward.

In the following we demonstrate that the algorithm in Eq. (3.48) can be further simplified for implementation by approximating the computation of the partial derivative term $\frac{\partial E_t(\mathbf{F})}{\partial \mathbf{f}_l}$. Recall that the term $E_t(\mathbf{F})$ uses MFD terms of the form $MFD(\mathbf{f}_{l'}, \mathbf{f}_l)$ to enforce temporal domain smoothness in the recovered video. According to its definition, each term $MFD(\mathbf{f}_{l'}, \mathbf{f}_l)$ essentially captures the motion compensated difference between the frame $\mathbf{f}_l$ and its reference frame $\mathbf{f}_{l'}$. Suppose that we approximate the term $MFD(\mathbf{f}_{l'}, \mathbf{f}_l)$ by replacing the reference frame $\mathbf{f}_{l'}$ by its most recently available estimate, say the most recent iterate $\mathbf{f}_{l'}^k$, i.e.,

$$MFD(\mathbf{f}_{l'}, \mathbf{f}_l) \approx MFD(\mathbf{f}_{l'}^k, \mathbf{f}_l). \tag{3.49}$$

Then $E_t(\mathbf{F})$ can be approximated by using the terms $MFD(\mathbf{f}_{l'}^k, \mathbf{f}_l)$. For example, the $E_t(\mathbf{F})$ defined in Eq. (3.17), can be approximated as

$$E_t(\mathbf{F}) \approx MFD(\mathbf{f}_1^k, \mathbf{f}_2) + MFD(\mathbf{f}_2^k, \mathbf{f}_3) + \cdots + MFD(\mathbf{f}_5^k, \mathbf{f}_6) + MFD(\mathbf{f}_6^k, \mathbf{f}_7). \tag{3.50}$$

With $E_t(\mathbf{F})$ approximated, the partial derivative $\frac{\partial E_t(\mathbf{F})}{\partial \mathbf{f}_l}$ can then be computed as

$$\frac{\partial E_t(\mathbf{F})}{\partial \mathbf{f}_l} \approx \sum_{l'} \frac{\partial}{\partial \mathbf{f}_l} MFD(\mathbf{f}_{l'}^k, \mathbf{f}_l) \triangleq \frac{\partial \hat{E}_t(\mathbf{F})}{\partial \mathbf{f}_l}, \tag{3.51}$$

where the summation is over the frames $\mathbf{f}_{l'}$ that are used to predict $\mathbf{f}_l$. Clearly, the evaluation of $\frac{\partial \hat{E}_t(\mathbf{F})}{\partial \mathbf{f}_l}$ requires only computations in the form of Eq. (3.40).

Replacing $\frac{\partial E_t(\mathbf{F}^k)}{\partial \mathbf{f}_l}$ by $\frac{\partial \hat{E}_t(\mathbf{F}^k)}{\partial \mathbf{f}_l}$ in Eq. (3.48), we obtain the following simplified recovery algorithm

$$\mathbf{f}_l^{k+1} = P_{C_l}\left[\mathbf{f}_l^k - \alpha\left(2(\mathbf{f}_l - \mathbf{g}_l) + \lambda_1 \frac{\partial}{\partial \mathbf{f}_l}E_s(\mathbf{F}^k) + \lambda_2 \frac{\partial}{\partial \mathbf{f}_l}\hat{E}_t(\mathbf{F}^k)\right)\right], \quad (3.52)$$

for $l = 1, 2, \cdots, L$. Clearly, the computations in Eq. (3.41) are no longer needed in this simplified algorithm, thereby the necessary bookkeeping associated with pixel prediction is bypassed.

Similar simplifications can be used when the definition in Eq. (3.20) is used for $E_t(\mathbf{F})$. Then, the motion compensated estimate of frame l using m as reference is approximated by

$$\hat{f}_m^{\,l}(i,j) \approx (f_m)^k(i + p^i, j + q^j) \quad (3.53)$$

where $(\mathbf{f}_m)^k$ is the estimate of $\mathbf{f}_m$ available from the most recent iteration k and (p^i, q^j) the appropriate motion vectors. In this case the dependency of $\hat{\mathbf{f}}_m^{\,l}$ on $\mathbf{f}_m$ is removed. Thus, in the calculation of the gradient $E_t(\mathbf{F})$ similar rules as above apply for the MFD part of it. Furthermore, for calculation of the gradient of the cross-correlation of the motion compensated errors calculation of the terms in Eq. (3.46) is not necessary. This as mentioned previously simplifies the computations and results not only in computational simplicity but also in significant memory savings making this algorithm more attractive for practical implementation.

3.5   DETERMINATION OF REGULARIZATION PARAMETERS

The goal of regularization in a recovery algorithm is to introduce prior knowledge, see for example [8]. In our case this is the temporal and spatial smoothness of the video sequence. As is easy to see from the objective function in (3.6) the role of the regularization parameters λ_1, λ_2 is to define the trade-off between the fidelity to the data and prior information. The selection of good values for the regularization parameters is a very important problem in image recovery. However, in traditional image recovery the original (source) images are not available. However, in our case the original images are available at the coder. Therefore, in what follows we propose an iterative algorithm that determines the values of the regularization parameters from the original images in the coder.

The regularization parameters λ_1, λ_2 are found iteratively by

$$(\lambda_1^{\,k}, \lambda_2^{\,k}) = arg \min_{\lambda_1, \lambda_2} ||\mathbf{F} - \mathbf{F}^k(\lambda_1^{\,k}, \lambda_2^{\,k})||^2 \quad (3.54)$$

where $\mathbf{F}$ the original images in the coder and

$$\mathbf{F}^{k+1}(\lambda_1^{\,k+1}, \lambda_2^{\,k+1}) = \left(P_C\mathbf{F}^k(\lambda_1^{\,k}, \lambda_2^{\,k}) - \alpha\nabla J(\lambda_1^{\,k+1}, \lambda_2^{\,k+1} P_C\mathbf{F}^k(\lambda_1^{\,k}, \lambda_2^{\,k}))\right), \quad (3.55)$$

where P_C is the projection operator onto the data set. The dependence of the function $J(\mathbf{F})$ on (λ_1, λ_2) is linear and is seen in Eq. (3.6). Therefore, it is easy to obtain in closed form $(\lambda_1{}^k, \lambda_2{}^k)$ from Eq. (3.54).

We observed that the iteration in (3.54), (3.55) converged and the termination of this iteration was determined using a threshold and the following criterion:

$$\epsilon = |\lambda_i^{k+1} - \lambda_i^k| \qquad \text{for } i = 1, 2. \tag{3.56}$$

with ϵ a given threshold. In our experiments $\epsilon = 10^{-4}$ was used for the given $\alpha = 0.01$.

The convergence properties of this iterative algorithm and the sensitivity of the recovered images to the choice of regularization parameters was examined in a sequence of numerical experiments that follows.

3.6 NUMERICAL EXPERIMENTS

Numerical experiments were performed to test the proposed video recovery algorithms. These algorithms where applied to H.261/3 compressed video streams, however, they can be applied to MPEG compressed video also. The "Mother and Daughter(MD)", "Foreman", "Carphone" and "Suzie" sequences were selected as test sequences. The processed picture format was QCIF(Y: 176 × 144, U and V: 88 × 72) for all test sequences. A total of 30 frames from each selected sequence were used in our experiments.

As an objective metric the Peak-Signal-to-Noise-Ratio $(PSNR)$ is used. This metric is defined by

$$PSNR = 10log_{10} \frac{255^2 \cdot W}{\|\mathbf{f} - \hat{\mathbf{f}}\|^2} \ dB, \tag{3.57}$$

where W is the number of pixels in the image, and $\mathbf{f}$ and $\hat{\mathbf{f}}$ are the original and the recovered image, respectively.

The regularization parameters λ_1, $\beta\lambda_1$ and λ_2 were estimated in the coder using the algorithms of Eqs. (3.54) and (3.55). For these experiments the value of α, the relaxation parameter was chosen to be 0.01. The estimated values of λ_1, $\beta\lambda_1$ and λ_2 of test video sequence were transmitted to the decoder and were used in the proposed recovery based multichannel iterative decoding process.

Experiment 1: For this experiment H.261 was used as the video coding algorithm and a number of different regularization approaches were tested. More specifically, the following cases were considered.

1. For this case only spatial regularization was used only. In other words the function $J(\mathbf{F})$ in Eq. (3.22) does not contain the $E_t(\mathbf{F})$ term. The purpose of this experiment was to demonstrate the performance improvement that temporal regularization introduces.

2. Temporal regularization was used and the transmitted motion vectors by the H.261 coder were used. However, for this case we made the assumption that the motion trajectories are linear and the velocity is constant for every three frames. Thus, if the forward motion field between frames $l-1$ and l is known the backward motion field between frames $l+1, l$ is also known. Clearly this is a simplifying assumption which is valid for translational motion. However, with this assumption it is possible to construct a bi-directional motion compensated estimate of frame l according to $\hat{\mathbf{f}}_l^{MC} = \frac{1}{2}(\hat{\mathbf{f}}_{l+1}^l + \hat{\mathbf{f}}_{l-1}^l)$ Then, temporal regularization according to Eq.(3.20) can be used. Furthermore, the approximation in Eq. (3.53) was used to simplify the implementation.

3. Temporal regularization was used again in a similar fashion as in the previous case. However, for this case the approximation in Eq. (3.53) was not used.

4. Temporal regularization as in Eq. (3.20) was used again. However, for this case a dense motion field on a pixel-by-pixel basis was estimated using a block-matching full search algorithm from a smoothed version of the compressed frames. More specifically, 5-channel regularization was used. In other words the motion compensated estimate of frame l is given by $\hat{\mathbf{f}}_l^{MC} = \frac{1}{4}(\hat{\mathbf{f}}_{l-2}^l + \hat{\mathbf{f}}_{l-1}^l + \hat{\mathbf{f}}_{l+1}^l + \hat{\mathbf{f}}_{l+2}^l)$. This motion compensated estimate requires four motion fields. This approach is very computationally intensive and to a certain extent impractical. However, it establishes an upper performance bound for this algorithm.

All four above described regularization approaches were applied to 30 frames of the MD, "Foreman", "Carphone" and "Suzie" sequences which were compressed at $23.77, 31.69, 28.25$ and 24.37 Kbps, respectively. The $PSNR$ results versus frame number are shown in Figure 3.1(a),(b),(c) and (d), respectively. To visually evaluate the results also the original, the compressed and the recovered images using the previously described regularization approaches are shown in Figure 3.2 (a),(b) and (c)-(f), respectively, for one frame of the "Foreman" sequence.

Experiment 2: For this experiment H.263 was applied with unrestricted motion vector mode (annex-D) advanced prediction mode (annex-F), PB frame mode (annex-G) and syntax based arithmetic coding (annex-E) [5]. Similar to experiment 1 five different types of multichannel regularization were tested.

1. Temporal regularization was used based only on the transmitted motion vectors by the H.263 coder.

2. Temporal regularization based only on the provided motion vectors by the H.263 decoder and the linear trajectory constant velocity motion assumption as in case 3 in experiment 1. For this case the approximations in Eq. (3.51) and (3.53) were not used.

3. Temporal regularization as previously with the linear trajectory constant velocity motion assumption. However, for this case the approximations in Eq. (3.51) and (3.53) were used.

4. Temporal 5-channel regularization using a dense motion field on a pixel-by-pixel basis like case 4 of Experiment 1.

The MD and the "Foreman" sequences were used to test the proposed recovery algorithms at compression rates 10.67 and 16.63 kbps, respectively. In Figure 3.3 (a) and (b) the PNSR versus frame number results are shown for the MD and "Foreman" sequence, respectively. In order to evaluate the visual results, in Figure 3.4 (a), (b) and (c)-(f) the original, compressed and recovered using the types of temporal regularization previously described frame #22 of the MD sequence are shown, respectively.

Experiment 3: The purpose of this experiment was to investigate the sensitivity and the properties of the proposed iterative algorithm for estimation of the regularization parameters. More specifically, the parameters $\lambda_1^*, \beta^* \lambda_1^*, \lambda_2^*$ are the values of $\lambda_1, \beta \lambda_1, \lambda_2$, that the algorithm in Eq. (3.54) and (3.54) converged to. Two experiments were performed. In the first, the proposed recovery algorithms were applied using $\gamma \lambda_1^*, \gamma \beta \lambda_1^*, \gamma \lambda_2^*$, where γ was a scaling parameter that ranged from 10^{-2} to 10^2. In the second, we used $\gamma \lambda_1^*, \gamma \beta^* \lambda_1^*$ and λ_2^*. This last experiment, tests the sensitivity of the recovery algorithm with respect to spatial vs. temporal regularization.

In Figure 3.5 we provide plots of the MSE of the recovered frame versus the scaling parameter γ for H.261 compressed sequences. In all cases 5-channel multichannel regularization was used using a recalculated dense motion field. From these plots it is clear that the proposed algorithm for selecting regularization parameters provides a good choice for these parameters. Furthermore, the proposed algorithms are quite robust to the choice of regularization parameter. In another set of experiments, that we do not include here, we found out that the projection onto the data set reduces significantly the sensitivity of this type of algorithms to the value of the regularization parameters.

From all the experiments presented in this chapter it is clear that the difference in visual results is not very noticeable for more complicated temporal regularization approaches. Furthermore, we observed that the proposed gradient-projection algorithm converged very rapidly in less than 10 iterations.

3.7 CONCLUSIONS

In this chapter we formulated the video decoding problem as a multichannel regularized recovery problem. Since in the video decoder an image has to be reconstructed from missing information regularization introduces prior knowledge that complements the transmitted data. The novelty of the proposed recovery algorithm is the use in addition to spatial, temporal regularization also. This is accomplished by enforcing smoothness along the motion trajectories which are known by the transmitted motion field.

We presented numerical experiments using H.261 and H.263 compressed image sequences. Our numerical experiments demonstrated that temporal regularization offers a significant advantage both in term of PSNR and visual effects. Furthermore, for H.263 since a denser motion field is transmitted the improvements were more significant.

References

[1] W. B. Pennebaker and J. L. Mitchel, *JPEG: Still Image Data Compression Standard,* Van Nostrand Reinhold, 1993.

[2] ISO/IEC 11172-2, "Information Technology – Coding of Moving Pictures and Associated Audio for Digital Storage Media up to about 1.5 Mbits/s – Video," Geneva, 1993.

[3] ISO/IEC 13818-2, "Information technology – Generic coding of moving pictures and associated audio: Video," Nov. 1994.

[4] ITU-T Recommendation H.261, *Video Codec for Audiovisual Services at $p \times 64$ kbits.*

[5] "ITU-T Recommendation H.263, *Video Coding for low bitrate communication*, Sept. 1997.

[6] Thomas Sikora, "MPEG digital video-coding standard," *IEEE Signal Processing Magazine,* Vol. 15, No. 5, pp.82–100, Sept. 1997.

[7] B. Ramamurthi and A. Gersho, "Nonlinear space-variant postprocessing of block coded images," *IEEE Trans. on Acoust., Speech and Signal Processing,* Vol. 34, No. 5, pp. 1258-1267, October 1986.

[8] N. P. Galatsanos and A. K. Katsaggelos, "Methods for choosing the regularization parameter and estimating the noise variance in image restoration and their relation", *IEEE Trans. on Image Processing,* Vol. 1, No. 3, pp. 322-336, J uly, 1992

[9] K. Sauer, "Enhançement of low bit-rate coded images using edge detection and estimation", *Computer Vision Graphics and Image Processing: Graphical Models and Image Processing*, Vol. 53, No. 1, pp. 52-62, January 1991.

[10] S. Minami and A. Zakhor, "An optimization approach for removing blocking effects in transform coding," *IEEE Trans on Circuits and Systems for Video Tech.*, Vol. 5, No. 2, pp. 74-82, April 1995.

[11] C. Kuo, and R. Hsieh, "Adaptive postprocessor for block encoded images," *IEEE Trans on Circuits and Systems for Video Tech.*, Vol. 5, No. 4, pp. 298-304, August 1995.

[12] Y. L. Lee, H. C. Kim, and H. W. Park, "Blocking effect reduction of JPEG images by signal adaptive filtering," *IEEE Trans. on Image Processing,*, vol. 7, no. 2, Feb. 1998.

[13] R. Rosenholtz and A. Zakhor, "Iterative procedures for reduction of blocking effects in transform image coding", *IEEE Trans on Circuits and Systems for Video Tech.*, Vol. 2, No. 1, pp. 91-94, March 1992.

[14] Mungi Choi and N. P. Galatsanos, "Multichannel regularized iterative restoration of motion compensated image sequences," *Journal of Visual Communication and Image Representation,* Vol. 7, No. 3, pp. 244-258, Sept. 1996.

[15] Y. Yang, N. Galatsanos and A. Katsaggelos, "Regularized reconstruction to reduce blocking artifacts of block discrete cosine transform compressed images," *IEEE Trans on Circuits and Sys. for Video Tech.*, Vol. 3, No. 6, pp. 421-432, Dec. 1993.

[16] Y. Yang, N. Galatsanos and A. Katsaggelos, "Projection-based spatially adaptive image reconstruction of block transform compressed images" *IEEE Trans. on Image Processing*, Vol. 4, No. 7, pp. 896-908, July 1995.

[17] T. Ozcelik, J. Brailean, and A. K. Katsaggelos, "Image and video compression algorithms based on recovery techniques using mean field annealing," *IEEE Proceedings*, Vol. 83, No. 2, pp. 304-316, February 1995.

[18] T. O'Rourke and R. Stevenson, "Improved image decompression for reduced transform coding artifacts," *IEEE Trans. on Circuits and Sys. for Video Tech.*, Vol. 4, No. 6, pp. 490-499, Dec. 1995.

[19] J. Luo, C. Chen, K. Parker, and T. S. Huang, "Artifact reduction in low bit rate DCT-based image compression," *IEEE Trans. on Image Processing*, vol. 5, pp. 1363–1368, Spet. 1996.

[20] Y. Yang and N. Galatsanos, "Removal of compression artifacts using projections onto convex sets and line process modeling", *IEEE Trans on Image Processing,* Vol. 6, No. 10, pp. 1345–1357, Oct. 1997.

[21] M. G. Choi, Y. Yang and N. P. Galatsanos, "Multichannel regularized recovery of compressed video," *Int. Conf. on Image Processing*, pp. 271-274, Santa Barbara, Nov. 1997.

[22] A. Tikhonov and V. Arsenin, *Solution of Ill-Posed Problems,* New York: Wiley, 1977.

[23] J. M. Ortega and W. C. Reinbolt, *Iterative Solutions to Nonlinear Equations in Several Variables,* New York, Academic, 1970.

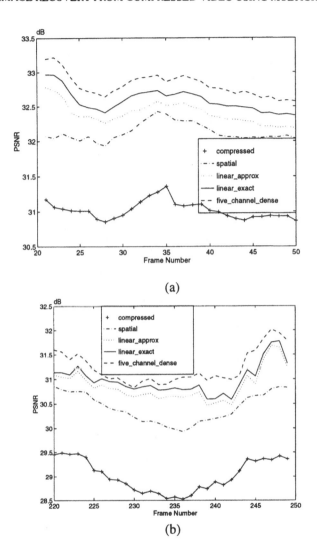

(a)

(b)

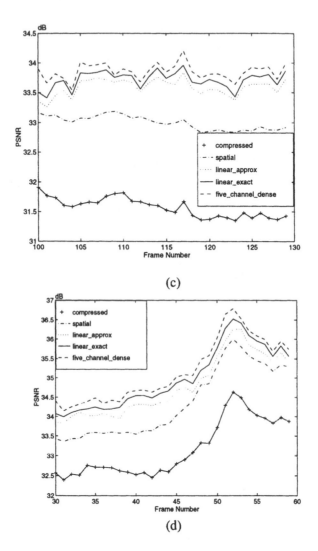

Figure 3.1 $PSNR$ v.s. # frame plots using different types of regularization for recovery. Results for (a) MD, (b) Foreman, (c) Carphone and (d) Suzie sequences are shown. H.261 was used for compression.

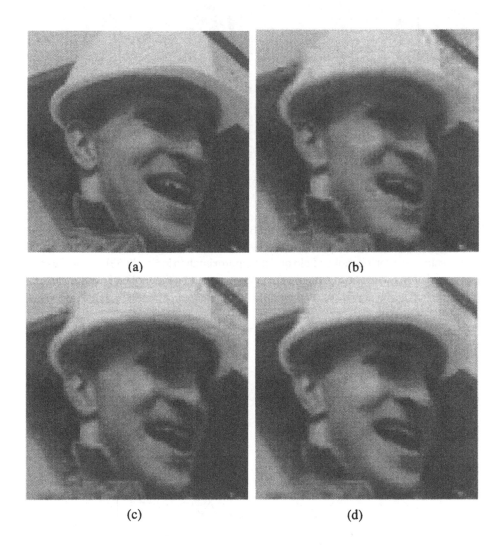

(a)   (b)

(c)   (d)

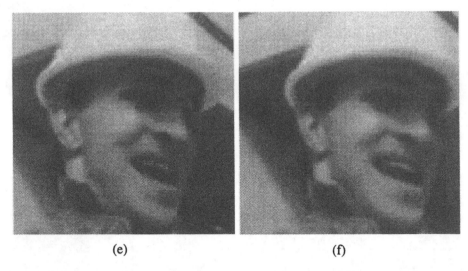

(e)   (f)

Figure 3.2 Foreman sequence, frame # 236, (a) original, (b) compressed, (c) recovered with spatial regularization only, (d) recovered with temporal regularization using the linear assumption for the motion field and the approximation in Eq. (3.53), (e) recovered with temporal regularization using the linear assumption for the motion field without the approximation in Eq. (3.53), (f) recovered using 5-channel temporal regularization and a dense motion field.

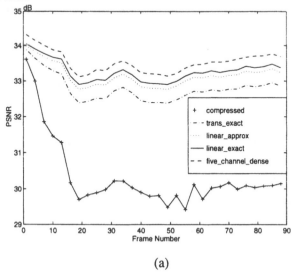

(a)

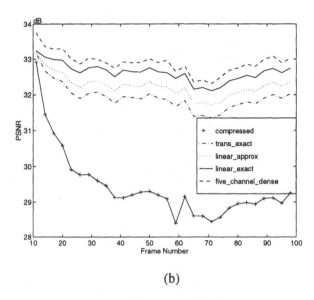

(b)

Figure 3.3 $PSNR$ v.s. # frame plots using different types of regularization for recovery. Results for (a) MD and (b) Foreman sequences are shown. H.263 was used for compression.

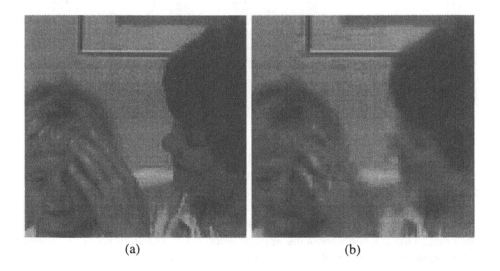

       (a)   (b)

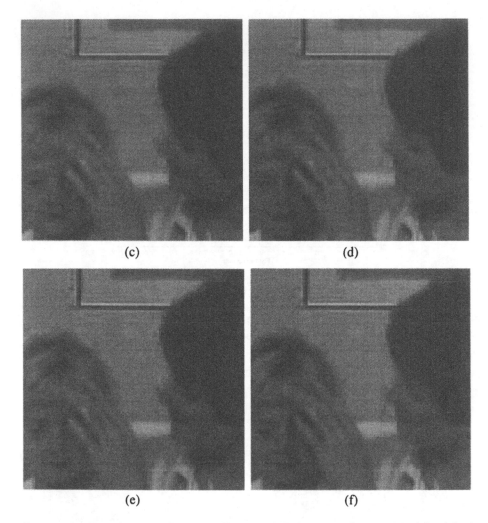

Figure 3.4 Sample image from experiment 2, MD sequence frame #22. (a) original, (b) compressed, (c) recovered with temporal regularization using the transmitted motion vectors only, (d) recovered with temporal regularization using the transmitted motion vectors and the linear motion assumption and the approximation in Eq. (3.53), (e) recovered with temporal regularization using the transmitted motion vectors and the linear motion assumption without the approximation in Eq. (3.53), (f) recovered using 5-channel temporal regularization and a dense motion field.

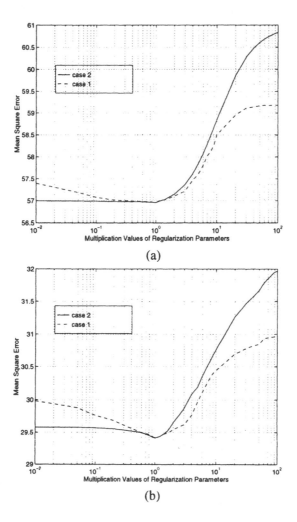

(a)

(b)

Figure 3.5   Regularization parameters sensitivity plots for H261, (a) 117th frame of the "Carphone" sequence (b) 237th frame of the "Foreman" sequence. For both cases 5-channel multichannel was used with a dense motion field for temporal regulatiation. The solid line shows the second case when only $(\lambda_1, \beta\lambda_2)$ are multiplied by γ.

4 SUBBAND CODED IMAGE RECONSTRUCTION FOR LOSSY PACKET NETWORKS

Sheila S. Hemami[1] and Robert M. Gray[2]

[1] School of Electrical Engineering
Cornell University
Ithaca, NY 14853
hemami@ee.cornell.edu

[2] Department of Electrical Engineering
Stanford University
Stanford, CA 94305
gray@isl.stanford.edu

Abstract:

An algorithm is presented for reconstructing image subband coefficients lost or delayed during transmission. The algorithm consists of two parts, allowing reconstruction of both low-frequency and high-frequency coefficients. Low-frequency coefficients are reconstructed using bicubic interpolation in which the interpolation grid is adapted to achieve accurate edge placement in the synthesized image. The adaptation is based on subband analysis of an edge model, which demonstrates that edges can be identified using the decimated high-frequency subbands without requiring edge detection on the low-frequency band itself. High-frequency coefficients are reconstructed using one-dimensional linear interpolation, which provides good visual performance as well as maintaining properties required for edge placement in the low-frequency reconstruction. The algorithm performs well on losses of single coefficients, 1-d vectors, and small blocks, and is therefore applicable to a variety of coding techniques.

4.1 INTRODUCTION

Digitally coded still images can be transmitted in raw format over a perfect communication channel, or with channel coding to compensate for individual bit errors. Packet-based transmission of image data introduces more variability into the transmission quality: networks can provide differing qualities-of-service with respect to packet loss or delay. In the case of real-time transmission, delay can be equivalent to loss. Transmission of digitally coded images over lossy packet networks presents a reconstruction problem at the decoder. Loss of coded data can produce catastrophic effects in the received images, and correction of errors caused by data loss is imperative to provide consumer-grade visual quality. Many techniques exist for transmitting raw data over such links (e.g., forward error correction (FEC), automatic retransmission query protocols (ARQ)) [1, 2]. However, unlike raw data, which must be received perfectly, visual data contains a great deal of redundancy which can be exploited to reconstruct the damaged image data while not providing an exact replica of the original. Providing that human perception is considered, the visual data can be successfully reconstructed using lossy signal processing techniques which exploit perceptual qualities and correlation within the signal. Ideally, these techniques should provide the highest quality reconstructed data possible with a minimal computational requirement. They can then be easily incorporated into existing systems, demanding minimal load from on-board video boards in workstations, or consuming minimal battery power in hand-held receivers, for example.

Error concealment for JPEG-coded data has been discussed previously, and this chapter now considers signal processing techniques for error concealment for subband and wavelet-coded images that have been hierarchically decomposed and intraband coded. When an image subband decomposition is intraband coded, the encoded stream provides spatial scalability, allowing images of varying spatial resolutions to be progressively decoded. In particular, a recovery technique for images coded with quadrature mirror filters (QMFs) is described. The technique is easily extended to any linear phase filters, including biorthogonal wavelet filters.

Because this work is concerned with signal processing techniques for error concealment, it is assumed that the locations of coefficients lost in transmission is known. This can be achieved through fixed-length coding combined with strategic packetization (for example, use of sequence numbers in packet headers, fixed-length packets, and a fixed data transmission order). While such a stragegy may not maximize compression performance, the resulting end-to-end system performance can be superior in a system where source encoding and packetization have been designed to facilitate error concealment when compared to a system with separately optimized source and channel coding. In particular, the described reconstruction technique does not limit the amount of lost data that can be reconstructed, and requires only a slight increase in data rate imposed by the coding and packetization constraints. In contrast, forward error correction (FEC) or retransmission limit the amount of data that can be reconstructed to at most the amount of overhead information, and possibly less, depending on spacings of lost data in case of FEC and transmission success in the case of retransmission. As such, the proposed reconstruction technique is more efficient in dealing with higher loss percentages.

In a hierarchically-decomposed subband coded image, the effects of loss in the high frequency subbands and the low frequency subband varies dramatically. Coefficient loss without error concealment in the high frequency bands may or may not be visible, depending on where the loss occurs. In contrast, because the low frequency band contains over 95% of the analyzed image signal energy and has a non-zero mean, lost coefficients and the resulting reconstruction errors can have a significant effect on the quality of the synthesized image. If no error concealment in the low frequency band is performed, dark holes appear in the synthesized image, spread out to an extent determined by the number of decomposition levels and the filter length. In any error concealment strategy for subband coded images, emphasis must be placed on providing high quality reconstruction of coefficients in the low frequency band. Humans are very sensitive to errors in continuous edges, so reconstruction of coefficients to maintain sharp edges and edge continuity is critical. Any algorithm for low frequency reconstruction must generate coefficients that maintain clean edge structures when the reconstructed coefficients are synthesized.

An approach to low-frequency coefficient error concealment for intraband-subband coded images was presented in [3]. The algorithm iteratively generates lost coefficients to minimize the mean squared error between the correctly received coefficients and the coefficients resulting from the analysis of the synthesized reconstructed image. This technique explots correlation between subbands at the same scale, and hence relies on the non-ideality of the analysis filters. As such, it is limited to systems in which the analysis filters have significant overlap in the passbands to provide high interband correlation. Good results were observed with 2-tap and 4-tap perfect reconstruction filters, while longer filters with sharper cutoffs proved unsuitable because of the decreasing correlation between subbands of differing orientation at the same scale.

In this chapter, a different approach is taken. Lost coefficients are interpolated using known coefficients within the same band. Correlation across the bands is exploited by adapting the interpolation of the low frequency coefficients based on high frequency coefficients at the same scale. In particular, the low frequency reconstruction algorithm exploits interband relationships and is based on three properties of the hierarchical subband decomposition: the lowest frequency band visually exhibits smoothness, this band has high horizontal and vertical correlation coefficients, and the high frequency bands at the lowest decomposition level contain horizontal and vertical edge information corresponding to the lowest frequency band. To maintain smoothness and exploit the high intraband correlation, a cubic interpolative surface is fit to known coefficients to interpolate lost coefficients. Accurate interpolation is achieved by adapting the interpolation grid in both the horizontal and vertical directions as determined by the high frequency bands.

High frequency reconstruction in bands that have been high-pass filtered in only one direction exploits correlation in the low-pass filtered direction. Linear interpolation provides good visual performance as well as maintains properties required for accurate interpolation in the low frequency reconstruction algorithm.

Section 4.2 introduces a simple edge model for edges in images. This edge model is analyzed with both even and odd length QMFs, and relationships are derived between subsampled low frequency and high frequency coefficients around edges. It is shown that the presence of edges and the subsampling pattern can be accurately identified with high probability by examining the high frequency subband coefficients.

Section 4.3 describes incorporation of the obtained edge information into bicubic interpolation for low frequency coefficients. The subsampling pattern information recovered from the high frequency coefficients allows accurate interpolation of low frequency coefficients based on an adaptation of the interpolation grid resulting from rectangular sampling.

Section 4.4 describes the use of one-dimensional linear interpolation for high frequency coefficient reconstruction in bands that have been high-pass filtered in a single direction, and finally Section 4.5 presents implementation aspects and reconstruction results.

4.2 AN EDGE MODEL FOR IMAGES AND ITS ANALYSIS

Reconstructing coefficients so that edges are accurately synthesized is crucial to providing visually acceptable images. The subband decomposition provides a natural framework through which relationships between low and high frequency subbands can be used to characterize the low frequency signal for accurate edge reconstruction. Commonly, describing interband relationships in subband decompositions has relied on heuristic techniques and additional processing of the low frequency band to relate activity in the low and high frequency signals. In [4], an empirically derived threshold measure is used to determine activity for each coefficient in the low frequency band, and this activity is used to predict the amplitudes of high frequency coefficients. In [5], an edge detector is applied to the low frequency band, the output of which is then thresholded. A window of three high frequency coefficients centered on locations above the threshold is then selected as significant in representing edge structures in synthesized images. In both cases, the low frequency coefficients are used to determine edge structures and hence to select important high frequency coefficients.

In the case of reconstruction, some low frequency coefficients are missing, so processing of the low frequency band cannot be used to accurately identify edge locations. However, the fact that edges are clearly visible in the high frequency bands suggests that they may be identified based on high frequency characteristics alone.

This section describes how high frequency subband behavior in the vicinity of edges is characterized and is used in reconstruction. A simple edge model is proposed and then analyzed using both high-pass and low-pass QMFs, indicating how to identify edges using only the high frequency coefficients, and how coefficients in the vicinity of edges should be reconstructed.

In the following, subbands in a hierarchical decomposition are referred to by two letters, corresponding to the last set of horizontal and vertical filters used to generate them, respectively. The low frequency subband is referred to as the *LL band*, indicating that the signal has been lowpass filtered in both the horizontal and vertical directions. Only one LL band exists, and it occurs in the *lowest decomposition level*. There are three types of high frequency subbands, the *LH, HL, and HH bands*. Multiple LH, HL, and HH bands exist, one set for each decomposition level. When these bands are discussed, the decomposition level is stated as either the lowest level, or as a *higher decomposition level*.

A simple one-dimensional edge model is defined and is used to illustrate the importance of correct edge placement for low frequency reconstruction. The model is defined as a three-valued signal,

$$(\ldots, p_1, p_1, mp_1 + (1 - m)p_2, p_2, p_2, \ldots), 0 < m \leq 1,$$

where either $p_1 > p_2$ or $p_1 < p_2$, and the mixture value is referred to as the *edge center*. Analysis of this signal yields two different low frequency subband signals, depending on whether the subsampling occurs in even or odd locations, as illustrated in Figure 4.1. For the example shown using a 5-tap QMF from [6], the even-subsampled low frequency signal resembles a step function, while the odd-subsampled signal resembles a ramp. Reconstruction of the second coefficient in the step incorrectly as a ramp results in the synthesized edge placement offset by one pixel to the left from its correct location.

The high frequency bands clearly contain edge information that can be selected as important without extra processing of the low frequency band to explicitly detect edges. Based on this observation, an analysis of the simple edge model using a high-pass QMF provides a quantitative characterization of the high frequency signal so that edge structures can be identified using the high frequency coefficients alone. An analysis of the edge model using a low-pass QMF then provides details about how low frequency coefficients in the vicinity of edges should be reconstructed. Edges are identified using the high frequency signal, and their relative locations to the lost low frequency coefficients determines the correct placement of synthesized edges in reconstruction.

In the following subsections, the three-valued edge model is analyzed using a high-pass quadrature mirror filter derived from a low-pass QMF, for both even and odd length filters, to identify characteristics in the high-frequency band that indicate the presence of an edge. While even and odd length filters have different coefficient

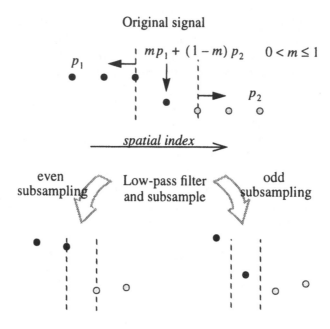

Figure 4.1 A simple edge model and its low frequency subband as a function of the subsampling position. The edge center value is synthesized at the location between the dashed lines. Reconstruction of the second coefficient in either signal incorrectly results in a misplacement of the synthesized edge.

relationships, it is shown that the end results of the analysis are independent of the filter type used, and edges can be easily identified by the presence of two large adjacent high-frequency coefficients. A corresponding analysis of the edge model using even and odd length low-pass QMFs provides the required information for reconstruction of low-frequency coefficients in the vicinity of edges. Use of the high frequency coefficients to determine edge placement information is then described.

In this section, the following conventions are used: the filter coefficient at $n = 0$ is underlined, and the subband coefficients are given as a function of k, representing the absolute distance between the edge center and the filter coefficient at $n = 0$. Coefficients *to the left of the edge* are coefficients corresponding to spatial indices less than the edge center index; coefficients *to the right of the edge* are coefficients corresponding to spatial indices greater than the edge center index. *Even subsampling* refers to subsampling when the coefficient at the edge center is discarded.

4.2.1 Analysis of the Edge Model with Even-length Filters

Consider a symmetric, even length (N) low pass filter, with one more causal tap than non-causal tap. This filter is represented by coefficients

$$\{c_{N/2-1} \cdots c_1 \underline{c_0}\, c_1 \cdots c_{N/2-1}\}.$$

Then the corresponding high-pass QMF is represented as

$$\{-c_{N/2-1} \cdots - c_1 \underline{c_0}\, - c_0\, c_1 \cdots c_{N/2-1}\}.$$

To analyze the behavior of the high frequency subband coefficients in the presence of edges, the edge model is filtered with the high-pass QMF and the resulting coefficients x are characterized in three locations. The high frequency subband coefficients to the left of the edge (prior to subsampling) are given by

$$x_{left}(k) = \left(\sum_{n=k}^{N/2-1} c_n\,(-1)^{n+1} \right) p_1 + \left(\sum_{n=k+1}^{N/2-1} c_n\,(-1)^n \right) p_2 \qquad (4.1)$$
$$+ (-1)^k\, c_k\,(m p_1 + (1 - m)p_2)$$

Equation (4.1) can be simplified to

$$x_{left}(k) = \left((-1)^k\, m c_k + \sum_{n=k}^{N/2-1} c_n\,(-1)^{n+1} \right) T_p \qquad (4.2)$$

where $T_p = (p_1 - p_2)$. In similar simplified form, coefficients to the right of the edge are given by

$$x_{right}(k) = \left((-1)^k\, m c_{k-1} + \sum_{n=k}^{N/2-1} c_n\,(-1)^{n+1} \right) T_p \qquad (4.3)$$

The high frequency coefficient at the edge center is

$$x_{center}(k) = \left(m c_0 + \sum_{n=0}^{N/2-1} c_n\,(-1)^{n+1} \right) T_p \qquad (4.4)$$

From Equations (4.2), (4.3), and (4.4), it is apparent that for a given even-length filter, the high frequency coefficients are functions of the coefficient location parameterized by $\{$left, center, right$\}$ and k, the edge parameter m, and the difference in pixel values across the edge T_p:

$$x_{loc}(k, m) = \alpha_{even}(k, m, loc) \, T_p \qquad (4.5)$$

Stronger edges have larger (in absolute value) high frequency coefficients, and the behavior of the coefficients in the vicinity of edges can be characterized as a function of m. Figure 4.2 plots $|x_{loc}(k, m)|$ with $T_p = 1$ for an 8-tap filter from [7].

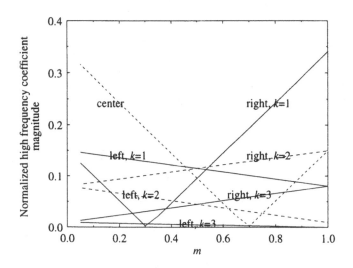

Figure 4.2 Normalized high frequency coefficient magnitude versus edge parameter m for an 8-tap filter. Solid lines represent even subsampling locations while dashed lines represent odd subsampling locations. The value k is the offset from the edge center.

Prior to subsampling, with the exception of small regions for m around 0.3 and 0.7, there are four adjacent filtered values that are at least an order of magnitude larger than surrounding values. These values occur immediately to the left of the edge, at the edge center, and at the two positions to the right of the edge. When the high frequency signal is subsampled, each of the even and the odd subsampling cases has two large adjacent coefficients in the vicinity of the edge, as illustrated in Figure 4.3, again with the exceptions of regions around $m = 0.3$ for even subsampling and $m = 0.7$ for odd subsampling (regions on the m axis where the two largest coefficients are not adjacent are referred to as *error regions*). By allowing an error for approximately 10% of edges with m in these ranges (assuming even and odd subsampling are equally likely), edges in the low frequency subband can be identified by examining the corresponding high frequency subband coefficients and the locations of large high frequency coefficients relative to the low frequency coefficient location. Similar results extend to the longer filters given in [7], with error occurring for only approximately 5% of the edges; the error region for even subsampling does not exist for filters of length 12 and higher, while the odd subsampling error region around $m = 0.7$ is present for most of the filters, an exception being filters of length 16.

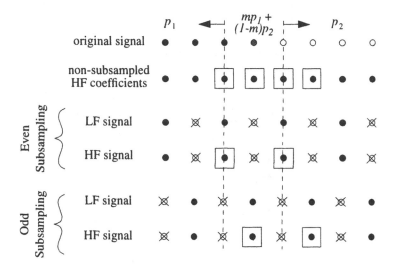

Figure 4.3   Locations of large high frequency coefficients and their positions relative to the edge center in both even and odd subsampled signals.

4.2.2   Analysis of the Edge Model for Odd-length Filters

Now consider an odd length filter, of length $N + 1$, N even. The low pass filter is represented by coefficients

$$\{c_{N/2} \cdots c_1 \underline{c_0} c_1 \cdots c_{N/2}\}.$$

The corresponding high pass filter is represented as

$$\{(-1)^{N/2-1}c_{N/2} \cdots \underline{c_1} - c_0 c_1 \cdots (-1)^{N/2-1}c_{N/2}\}$$

where, using the filter relationships for odd-length QMFs, the filter is delayed by one sample relative to the low-pass filter.

Coefficients to the left of the edge are given by

$$x_{left}(k) = \left(-c_0 + 2\sum_{n=1}^{k} c_n (-1)^{n+1} + \sum_{n=k+1}^{N/2} c_n (-1)^{n+1}\right) p_1 \quad (4.6)$$

$$+ \left(\sum_{n=k+1}^{N/2} c_n (-1)^{n+1}\right) p_2 + m\, c_{k+1}\, T_p\, (-1)^{k+1}$$

Coefficients to the right of the edge have values

$$x_{right}(k) = \left(-c_0 + 2\sum_{n=1}^{k-1} c_n (-1)^{n+1} + \sum_{n=k}^{N/2} c_n (-1)^{n+1}\right) p_2 \quad (4.7)$$

$$+ \left(\sum_{n=k}^{N/2} c_n (-1)^{n+1}\right) p_1 + m\, c_{k-1}\, T_p\, (-1)^{k}$$

The coefficient at the edge center is given by

$$x_{right}(k) = \left(\sum_{n=0}^{N/2} c_n (-1)^{n+1}\right) p_1 + \left(\sum_{n=1}^{N/2} c_n (-1)^{n+1}\right) p_2 \quad (4.8)$$

$$+ m\, c_1\, T_p$$

Because the signs of coefficients on either side of the symmetry point are the same, cancellation as in the case with even length filters does not occur, and at first glance, the equations do not seem to simplify nicely. However, the fact that the high-pass filter has zero DC response implies that

$$-c_0 + 2\sum_{n=1}^{N/2} c_n (-1)^{n+1} = 0 \quad (4.9)$$

$$= -c_0 + 2\sum_{n=1}^{j} c_n(-1)^{n+1} + 2\sum_{n=j+1}^{N/2} c_n (-1)^{n+1}$$

$$\forall j, 1 \leq j \leq (N/2 - 1).$$

Comparison of Equation (4.9) to the multiplication factors of p_1 and p_2 in Equations (4.6), (4.7), and (4.8) reveals that the coefficients of p_1 and p_2 are equal in magnitude and opposite in sign. Therefore, again for each filter the subband coefficient behavior can be characterized as a function of the coefficient location, the edge parameter, and the pixel difference across the edge

$$x_{loc}(k, m) = \alpha_{odd}(k, m, loc)\, T_p \qquad (4.10)$$

Figure 4.4 plots $|x_{loc}(k, m)|$ when $T_p = 1$ for a 5-tap filter from [6]. Examination of the non-subsampled high frequency signal shows that as in the case of the even-length filters, an edge is characterized by two large adjacent coefficients, with the exception of a region for even subsampling around $m = 0.5$. Allowing this error, even and odd subsampling patterns yield the same results diagrammed in Figure 4.3. As filter length increases for the longer odd length filters given in [6], the even subsampling error region remains around $m = 0.5$ and does not increase in size, while an error region for odd subsampling appears around $m = 0.9$ and has width of approximately 0.1.

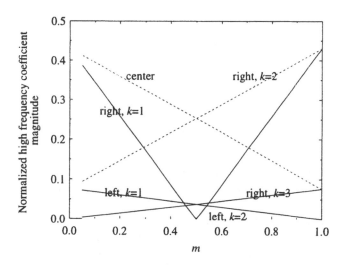

Figure 4.4   Normalized high frequency coefficient magnitude versus edge parameter m for a 5-tap filter. Solid lines represent even subsampling locations while dashed lines represent odd subsampling locations. The value k is the offset from the edge center.

4.2.3   Low Frequency Edge Model Analysis

The high frequency edge model analyses of the previous two subsections demonstrate that two adjacent large high frequency coefficients indicate the presence of a simple edge. A low frequency edge model analysis indicates how lost coefficients in the vicinity of an edge should be reconstructed. If a coefficient is weighted toward p_1, it should be reconstructed with more emphasis on the adjacent coefficient that has a higher weighting of p_1.

The edge model can be analyzed using both even and odd length low-pass QMFs, with the filters described in the previous two subsections. For example, for an even length low pass QMF, low frequency coefficients to the left of the edge center (prior to subsampling) are given by

$$
x_{left}(k) = \left(2 \sum_{n=0}^{k-1} c_n + \sum_{n=k}^{N/2-1} c_n + mc_k \right) p_1 + \left(\sum_{n=k}^{N/2-1} c_n - mc_k \right) p_2. \quad (4.11)
$$

Similarly, expressions can be written for the center coefficient and coefficients to the right of the edge center, and for analysis with an odd length low-pass QMF.

Each low frequency coefficient has the form of Equation (4.11); that is, each coefficient can be written as

$$
x = \xi_1 p_1 + \xi_2 p_2, \quad (4.12)
$$

$$
\text{where } \begin{cases} \xi_1 = a_1 + bm \\ \xi_2 = a_2 - bm \end{cases}
$$

The variables a_1 and a_2 are sums of filter coefficients and b is a single coefficient, and the relative magnitudes of ξ_1 and ξ_2 depend on the location of the low frequency coefficient with respect to the edge center, and in some cases, on the value of the edge parameter m. These values were evaluated for several even and odd length QMFs as given in [6, 7]. The results hold with any QMFs, due to the required characteristics of such filters.

First consider an even length QMF. For coefficients to the left of the edge with $k \geq 1, \xi_1 \gg \xi_2$, regardless of the value of m, because $a_1 \gg a_2$ and $b \ll a_1$. Likewise, for coefficients to the right of the edge with $k \geq 2, \xi_2 \gg \xi_1$, regardless of the value of m, because $a_2 \gg a_1$ and $b \ll a_2$ (the relation $\xi_2 \gg \xi_1$ follows directly from the result to the left of the edge with $k \geq 1$ because of the symmetry of the filter; however, the b values differ to the left and to the right of the edge). However, for the center value, $a_1 = a_2 = \sum_{n=0}^{N/2-1} c_n$, and $b = c_0 \approx a_1$. Therefore, the value of m strongly influences the center coefficient. Likewise, for the coefficient immediately to the right of the edge with $k = 1, a_2 \gg a_1$ but $b \approx a_2/2$, and again the coefficient is strongly influenced by m.

For odd length QMFs, similar results are obtained. For coefficients to the left of the edge with $k \geq 1, \xi_1 \gg \xi_2$, and for coefficients to the right of the edge with $k \geq 1, \xi_2 \gg \xi_1$. Only the center coefficient is a strong function of m, as $a_2 \gg a_1$ but

$b > a_2/2$. The results for both even and odd length filters are summarized in Table 4.1. These results are used in determining reconstruction parameters for low frequency coefficients.

Table 4.1 Edge content of low frequency subband coefficients.

Coefficient location with respect to edge center		$x = \xi_1 p_1 + \xi_2 p_2$	
		Even length QMF	Odd length QMF
left	$k \geq 1$	$\xi_1 \gg \xi_2$	
	center	$(\xi_1, \xi_2) = f(m)$	
right	$k = 1$	$(\xi_1, \xi_2) = f(m)$	$\xi_2 \gg \xi_1$
	$k \geq 2$	$\xi_2 \gg \xi_1$	

4.2.4 Edge Classification Using the Edge Model

Given that there are two large coefficients in the vicinity of edges, the edge classification for a single low frequency coefficient is determined by examining a window of five high frequency coefficients centered at the location of the lost low frequency coefficient, and looking for two adjacent large absolute values.

If there are zero, one, or two non-adjacent large coefficients, the edge classification is "normal," indicating that no edge is present. If there are two adjacent large coefficients, one of four patterns occur. If the pattern is $\{l\ l\ s\ s\ s\}$ or $\{s\ s\ s\ l\ l\}$ (s referring to small and l referring to large), then the lost coefficient is to the right or left of an edge, respectively. If the pattern is $\{s\ l\ l\ s\ s\}$ or $\{s\ s\ l\ l\ s\}$, then the coefficient is immediately on an edge or just beyond it, depending on the subsampling pattern. If there are more than two large coefficients occurring in any pattern, then the edge classification is "high frequency variations" (HFV), indicating that there is more high frequency activity than the simple edge model can predict. These five cases and the low frequency coefficient behavior summarized in Table 4.1 are used in determining interpolation points when reconstructing lost coefficients.

Use of the edge model is easily extended to two dimensions for both vertical and horizontal edge identification in the LL band by applying it independently in each direction. Because the LL band contains a signal low-pass filtered and subsampled in both directions, the original properties of the signal are generally maintained following one-dimensional analysis in either direction and the edge model applies independently in both the horizontal and vertical directions. The LH band at the lowest decomposition level contains the required high frequency information in the vertical direction to classify vertical edges, while the HL band contains the corresponding information in the horizontal direction.

In independently classifying both horizontal and vertical edges, diagonal edges are implicitly identified and classified. Assuming that a diagonal edge follows the edge model, low-pass filtering in one dimension yields a signal approximating the edge model as given in Table 4.1 — each low-pass filtered and subsampled row or column now resembles a noisy edge model. Filtered coefficients are predominantly a function of p_1, p_2, or are $f(m)$. Because these coefficients, when observed in the orthogonal direction prior to high-pass filtering, still resemble the edge model, the high-pass filtered signal indicates an edge in both the horizontal and vertical directions, as expected for a diagonal edge. This result provides the desired diagonal edge reconstruction.

4.2.5 Edge Identification and Classification Performance

To evaluate the usefulness of the edge model analysis for edge identification and classification, edges were identified in original images prior to subband analysis, and then the corresponding horizontal and vertical high-frequency classifications were determined using the HL and LH subbands, respectively. Evaluation therefore consists of two parts: accurate edge identification in the original image, and subsequent determination of high-frequency classification.

Edges were identified using pixels in the original image to the left (top) of and to the right (bottom) of the coefficient subsampling points. Define $p(n)$ as a pixel at a subsampling point. Then $\mathbf{p}_{left} = [p(n-3)\ p(n-2)\ p(n-1)]$ and $\mathbf{p}_{right} = [p(n+1)\ p(n+2)\ p(n+3)]$. Define the operators $mean(\cdot)$ and $stddev(\cdot)$ as the mean and standard deviation, respectively, of the elements in the argument vector. If $mean(\mathbf{p}_{left}) > mean(\mathbf{p}_{right})$, an edge matching the edge model is identified if

1. $mean(\mathbf{p}_{left}) \geq p(n) > mean(\mathbf{p}_{right})$

2. $mean(\mathbf{p}_{left}) - stddev(\mathbf{p}_{left}) > mean(\mathbf{p}_{right}) + stddev(\mathbf{p}_{right})$, and

3. $|mean(\mathbf{p}_{left}) - mean(\mathbf{p}_{left})| \geq T_p$

The second condition allows variation from the average value, which occurs in natural images, but restricts the variation so that the edge model is still roughly met. The third condition ensures that only edges with a transition greater than the pixel threshold T_p are selected. If $mean(\mathbf{p}_{left}) < mean(\mathbf{p}_{right})$, then the inequalities in the above conditions are reversed.

The threshold for identification of large coefficients in the LH and HL subbands, T_c, was determined by examining the classifications for edges identified using the above technique with varying T_c. For each edge, the horizontal and vertical classifications were determined as "normal," "edges," or "HFV," corresponding to less than 2, 2, and greater than 2 adjacent coefficients exceeding the threshold. In the event that both horizontal and vertical classifications were HFV, the threshold was increased until at least one directions yielded a non-HFV classification. Sample results for vertical edges in the *couple* image for both the 5-tap and 8-tap filters are given in Tables 4.2 and 4.3, respectively. Because only high-frequency coefficients were examined for pixels already identified as edges, any normal classifications are considered erroneous. Because HFV was not allowed in both directions, HFV classifications do not contribute to reconstruction errors.

For all edge thresholds T_p, as the coefficient threshold T_c increases, the numbers of normal classifications increase, while the numbers of edge and HFV classifications decrease because the number of coefficients exceeding the threshold T_c decreases. For a fixed coefficient threshold T_c, the ratio of percentages of edges to percentages of normal classifications (edges:normal) decreases as the edge threshold T_p increases. As T_p increases, the size of the large high frequency coefficients increases, making them larger and hence easier to identify. Because the normal classifications are considered erroneous, the coefficient threshold that maximizes the edges:normal ratio will minimize reconstruction errors caused by misclassification. The results given in Tables 4.2 and 4.3 indicate that selecting $T_c = 2$ maximizes this ratio for all edge thresholds given. Results for horizontal edges yield the same threshold. For $T_c = 2$, 35–50% of the edges are be reconstructed one-dimensionally in the opposite direction. Of the remaining edges, approximately 30–40% are classified as edges and 10–20% are misclassified as normal. However, the misclassified edges are not guaranteed to be incorrectly reconstructed. Because the energy in the QMFs is $\sqrt{2}$, the threshold will increase by a factor of 2 for each decomposition level.

Table 4.2   Percentages of classified vertical edges in *couple* as T_p and T_c vary, 4-band decomposition with 5-tap filter. $T_c = 2$ yields the highest edges:normal classification ratio for all edge transitions T_p.

T_c	Percentage of vertical edges classified as (edges, normal, HFV); 5-tap filter		
	$T_p = 5$	$T_p = 15$	$T_p = 25$
1	36, 20, 44	33, 16, 51	30, 15, 55
2	43, 21, 36	41, 15, 44	38, 12, 49
3	39, 41, 20	45, 27, 28	46, 20, 34
6	32, 56, 12	41, 41, 18	46, 31, 23
8	26, 66, 8	36, 52, 12	43, 41, 16
10	22, 73, 5	31, 61, 7	39, 50, 11

Table 4.3   Percentages of classified vertical edges in *couple* as T_p and T_c vary, 4-band decomposition with 8-tap filter. $T_c = 2$ yields the highest edges:normal classification ratio for all edge transitions T_p.

T_c	Percentage of vertical edges classified as (edges, normal, HFV); 5-tap filter		
	$T_p = 5$	$T_p = 15$	$T_p = 25$
1	36, 20, 44	32, 17, 51	29, 16, 55
2	43, 23, 35	41, 17, 43	38, 13, 48
3	37, 44, 19	43, 31, 26	43, 24, 32
6	29, 60, 11	37, 46, 17	41, 37, 22
8	22, 71, 7	30, 59, 11	36, 50, 14
10	17, 79, 4	24, 69, 7	30, 60, 10

4.3   EDGE-MODEL BASED SURFACE GENERATION FOR LOW FREQUENCY RECONSTRUCTION

The high correlation present in the low frequency subband suggests that lost coefficients can be reconstructed using their neighbors, and the smooth, natural appearance of this subband suggests that interpolation should maintain this smoothness. Bicubic interpolation is selected as the surface generation technique, and is then modified to incorporate the edge information from the high frequency subbands. The bicubic surface can be considered to be a cubic spline surface with first order continuity on the edges. Requiring only continuity in the first derivatives is better suited to interpolation of low frequency subband coefficients than imposing higher order continuity constraints. First derivatives can be reasonably estimated using differences of adjacent coefficients. Generation of higher order derivatives involves using more coefficients in a larger area, thus incorporating more global signal characteristics rather than local characteristics.

In bicubic interpolation, a surface cubic in both x and y is generated by specifying four points $f(x, y)$ on the corners of a grid (referred to as the *grid corners*) and their corresponding gradients $\partial f / \partial x$ and $\partial f / \partial y$, and cross derivatives $\partial^2 f / (\partial x \partial y)$. Subband coefficients generated by a two-dimensional separable filter lie on a regular sampling grid, so bicubic interpolation can be applied to reconstructing a lost coefficient by using the four corner coefficients as grid corners $f(x, y)$ and by using adjacent coefficients to calculate the required gradients using one- or two- sided differencing. The direct application of bicubic interpolation to reconstruct a single lost coefficient at the grid point (n, m) fixes the interpolation point at $(x, y) = (1/2, 1/2)$, and the reconstructed coefficient is given by

$$\hat{f}\left(\frac{1}{2},\frac{1}{2}\right) = \frac{1}{4} \times \mathbf{1}' \begin{bmatrix} f(n-1,m-1) \\ f(n+1,m-1) \\ f(n-1,m+1) \\ f(n+1,m+1) \end{bmatrix} + \frac{1}{16} \times \mathbf{1}' \begin{bmatrix} \partial f/\partial x|_{n-1,m-1} \\ -\partial f/\partial x|_{n+1,m-1} \\ \partial f/\partial x|_{n-1,m+1} \\ -\partial f/\partial x|_{n+1,m+1} \end{bmatrix} \quad (4.13)$$

$$+ \frac{1}{16} \times \mathbf{1}' \begin{bmatrix} \partial f/\partial y|_{n-1,m-1} \\ \partial f/\partial y|_{n+1,m-1} \\ -\partial f/\partial y|_{n-1,m+1} \\ -\partial f/\partial y|_{n+1,m+1} \end{bmatrix} + \frac{1}{64} \times \mathbf{1}' \begin{bmatrix} \partial^2 f/\partial x \partial y|_{n-1,m-1} \\ -\partial^2 f/\partial x \partial y|_{n+1,m-1} \\ -\partial^2 f/\partial x \partial y|_{n-1,m+1} \\ \partial^2 f/\partial x \partial y|_{n+1,m+1} \end{bmatrix}$$

where $\mathbf{1} = [1\ 1\ 1\ 1]'$. An illustration of the coefficient grid and the coefficients used to compute the quantities in equation (4.13) is given in Figure 4.5. The reconstructed value can be considered to be the mean of the four corners with correction factors. While inclusion of the derivative terms provides some incorporation of the local surface structure, it is not enough to accurately place edges as discussed earlier. A visual example of bicubic interpolation and the edge defects it causes is given in Figure 4.8. Hence the edge model and LH and HL information at the lowest decomposition level is used to adapt the otherwise regular interpolation grid to better reconstruct the edges.

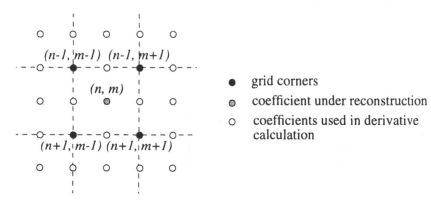

Figure 4.5 Subband coefficient sampling grid and coefficients used to interpolate the lost coefficient.

4.3.1 Grid Adaptation

One dimensional grid adaptation is conceptually understood as illustrated in Figure 4.6. First, a cubic polynomial is fit to two points and their derivatives on a regularly spaced grid, yielding an equation for the curve $\hat{f}(x)$. Interpolation at the center point

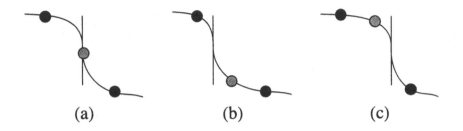

$$(a) \qquad\qquad (b) \qquad\qquad (c)$$

Figure 4.6 An illustration of grid adaptation. The three curves have identical end points, but different center values. The interpolated value $\hat{f}(1/2)$ occurs on the vertical line in each edge, and in the center in (a). By pushing this value to the left, in (b), or to the right, in (c), the interpolated value is biased toward the value on the right or left, respectively.

on the grid gives $\hat{f}(1/2)$ and places the point approximately between the two end points. However, adapting the grid by compressing it toward the left pushes the edge to the left and interpolates a value closer to the rightmost value in the center. This is mathematically equivalent to interpolating the desired point as $\hat{f}(z), z > 1/2$, where the value of z determines the extent of the compression. The interpolated point can be similarly evaluated to be closer to the leftmost value by evaluating $\hat{f}(z), z < 1/2$.

Adaptation is determined by the edge classification based on a window of five high frequency coefficients centered on the location of the lost LL coefficient as discussed in Section 4.2.4. The edge classification indicates the location of the reconstructed coefficient with respect to the edge, but in practice, it is too difficult to select an individual interpolation evaluation point for each reconstructed coefficient. Therefore, the interpolation points are quantized to $(1/4, 1/2, 3/4)$ which roughly correspond to compressing the interpolation grid to the right (Figure 4.6(c)), center-based interpolation (Figure 4.6(a)), and compressing the interpolation grid to the left (Figure 4.6(b)), and an appropriate quantized interpolation location is selected.

The quantized interpolation locations for each edge classification are determined using the low-pass analysis of the edge model outlined in Section 4.2.3. For example, consider even length filters. For odd subsampling, the low frequency coefficients selected are at $k = 2, 4, \ldots$ to the left of the edge, the edge center, and at $k = 2, 4, \ldots$ to the right of the edge. For the class $\{s \; l \; l \; s \; s\}$, the lost coefficient is located at $k = 2$ to the right of the edge. Based on the low frequency characterization in Table 4.1, this coefficient has more p_2 content, and should be reconstructed with a weighting toward the adjacent coefficient with greater p_2 content; that is, a weighting toward the coefficient on the right. Therefore, the interpolation location is $3/4$. The interpolation locations for the four edge classifications are summarized in Table 4.4. For the classes

$\{l\,l\,s\,s\,s\}$ and $\{s\,s\,s\,l\,l\}$, the interpolation location is the same for both even and odd subsampling. However, for $\{s\,l\,l\,s\,s\}$ and $\{s\,s\,l\,l\,s\}$, one of the locations is a function of m. Therefore, more information is required than simply the edge classification. Referring to the two large coefficients in the high frequency signal as l_{left} and l_{right}, this additional information is obtained by examining the ratio l_{left}/l_{right}, which is only a function of m for a given filter. The ratio behavior is filter dependent and is characterized by plotting the function versus m for each filter with even and odd subsampling. Interpolation points are then defined for various ranges of the ratio. This procedure can be simplified to simply selecting 3/4 for $\{s\,l\,l\,s\,s\}$ and 1/4 for $\{s\,s\,l\,l\,s\}$, in which case the location is correct for one subsampling pattern all the time, and for the other subsampling pattern half of the time.

Similar results hold for odd-length filters, except that only the center value is a function of m, corresponding to the $\{s\,s\,l\,l\,s\}$ class in odd subsampling. In this case, the ratio must be used to determine the interpolation point.

Table 4.4   Low frequency interpolation point as a function of the high frequency classification for even-length filters.

High frequency classification	Interpolation Location	
	Odd subsampling	Even subsampling
< 2 large adjacent values (normal)	1/2	
$\{l\,l\,s\,s\,s\}$	1/2	1/2
$\{s\,l\,l\,s\,s\}$	3/4	$f(l_{left}/l_{right})$ (1/2 or 3/4)
$\{s\,s\,l\,l\,s\}$	$f(l_{left}/l_{right})$ (1/4 or 1/2)	1/2
$\{s\,s\,s\,l\,l\}$	1/4	1/4
> 2 large adjacent values (HFV)	none	

4.3.2   Low Frequency Reconstruction Algorithm

The complete low frequency reconstruction algorithm is as follows. For each lost coefficient, the horizontal and vertical edge classifications are determined from the HL and LH subbands at the same scale by thresholding the absolute values of the subband coefficients. Three scenarios can occur. In the first, neither class is HFV. Then the interpolation locations are determined from the classifications, derivatives are estimated using one- or two-sided differencing, and unknown values that are required for calculation of either the four interpolation points or derivatives are estimated using weighted means, where unknown coefficients are estimated as the mean of available coefficients directly above, below, to the left, and to the right that are not classified as part of an edge structure based on their LH and HL classifications. The lost coefficient is then generated using edge-model based surface generation.

In the second scenario, the high frequency variations case is detected in only one direction. The lost coefficient is reconstructed following the procedure outlined above, but one-dimensional interpolation is performed in the direction with the non-HFV classification.

Finally, HFV can be detected in both directions. In this case, the threshold is increased until one direction has a non-HFV class and reconstruction proceeds as in the second scenario.

4.3.3 Extension to Adjacent Lost Coefficients

Because bicubic interpolation uses points not in the row or column of the lost coefficient, it is well suited to reconstructing adjacent coefficient loss. If multiple coefficients in a row or column are lost, the high frequency classification of the lost coefficients is examined and HFV-classified coefficients in the direction of loss are reconstructed one-dimensionally first. Following this one- dimensional reconstruction, coefficients are reconstructed using two-dimensional interpolation. Coefficients exhibiting high frequency variations in the direction perpendicular to the direction of loss are reconstructed last.

If the coefficients are coded in vectors of size $N \times 1$ or $1 \times N$, in the event of adjacent vector loss, the dimensions of a lost region will exceed one in both directions for some regions. In this case, grid adaptation is extended across two adjacent coefficients by modifying the interpolation location and high frequency classification for each of the adjacent coefficients. Figure 4.7 illustrates the coefficient grid and the coefficients used to compute the derivatives for adjacent coefficient reconstruction.The interpolation locations are quantized to $(1/6, 1/3, 1/2, 2/3, 5/6)$, to allow placement of a single edge between either of the two lost coefficients. As in grid adaptation for one coefficient, the possible high frequency patterns across two coefficients are enumerated and based on examination of the corresponding coefficient locations with respect to the edge, an interpolation point is assigned for each coefficient. These are enumerated in Table 4.3. Patterns not listed are either not possible (e.g. $\{l\ l\ s\ s\ l\ l\}$ produces a HFV classification for both coefficients) or not reconstructed using interpolation across two coefficients because at least one of the two classifications is HFV.

4.4 LINEAR INTERPOLATION FOR HIGH FREQUENCY RECONSTRUCTION

Random loss with no reconstruction in high frequency bands is typically visually imperceptible in natural images. However, because low frequency reconstruction relies on high frequency bands at the lowest decomposition level to place edges, accurate

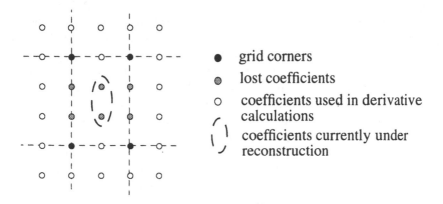

Figure 4.7   Adjacent coefficient reconstruction.

Table 4.5   Interpolation points for reconstructing two adjacent coefficients.

Left coefficient classification	Right coefficient classification	Left coefficient interpolation point	Right coefficient interpolation point
normal	normal	1/3	2/3
	$\{s\,s\,s\,l\,l\}$	1/6	1/3
$\{l\,l\,s\,s\,s\}$	normal	1/3	2/3
$\{s\,l\,l\,s\,s\}$	$\{l\,l\,s\,s\,s\}$	2/3	5/6
$\{s\,s\,l\,l\,s\}$	$\{s\,l\,l\,s\,s\}$	$f(l_{left}/l_{right})$	$f(l_{left}/l_{right})$
$\{s\,s\,s\,l\,l\}$	$\{s\,s\,l\,l\,s\}$	$f(l_{left}/l_{right})$	$f(l_{left}/l_{right})$

reconstruction of these bands is important. High frequency reconstruction is based on the fact that the HL and LH bands exhibit high correlation in the direction that has been low-pass filtered. A parallel exists between the LH/ HL and HH bands, as exists between the LL and LH/HL bands; that is, the same subband intermediate signal has been low-pass and high-pass filtered in one dimension. However, forming a model and characterizing its behavior is difficult because the intermediate signal has been high- pass filtered before the final low-pass/high-pass filtering stages. This high-pass filtering removes correlation and visual smoothness from the signal.

Fortunately, the high frequency signals can be adequately and accurately reconstructed using one-dimensional linear interpolation in the low frequency direction. Use of averaging is motivated by the results of linear minimum mean-square estimation (LMMSE) [8]. Unknown coefficients are interpolated from the two coefficients on either side of them in the low-pass direction. A single lost coefficient is reconstructed as $\hat{x}(n) = \alpha(x(n-1) + x(n+1))$, where $\alpha = \rho/(\rho^2 + 1)$ is the LMMSE solution and ρ is the one-dimensional, one-step correlation coefficient. Note that ρ must be only greater than $1/2$ to produce $\alpha > 0.4$. Assuming that this will generally be the case, the interpolation is simplified to $\alpha = 0.5$, for averaging.

Table 4.6 lists α values computed using measured correlation coefficients for two images with one and two decomposition levels. The *couple* image exhibits high low-pass direction correlation and hence the α values are close to the linear interpolation value. The *lena* image's HL bands are farther off, while the LH bands exhibit very low correlation even in the low-pass direction and hence the values are quite different. However, the correlation values used to calculate the α values given in the table are taken over the entire subbands, which are not spatially stationary. Calculating coefficients for individual rows and columns for the LH and HL bands, respectively, yields values closer to $\alpha = 0.5$, and experimental results demonstrate that for small numbers of adjacent lost coefficients, simple averaging provides adequate reconstruction.

The HH band exhibits low correlation in both horizontal and vertical directions. Lost coefficients are not reconstructed; they are set to zero.

Table 4.6 LMMSE interpolation α values for two images and one and two decomposition levels.

Image	Band	LMMSE α value, 1 lost coefficient
couple, 1-level	LH	0.44
	HL	0.47
couple, 2-levels	LH	0.44
	HL	0.44
lena, 1-level	LH	0.11
	HL	0.39
lena, 2-levels	LH	-0.00055
	HL	0.31

4.5 IMPLEMENTATION ASPECTS AND RECONSTRUCTION RESULTS

This section first describes system-level packetization and coding requirements to facilitate subband coefficient reconstruction using the described algorithm. While compression efficiency may suffer slightly, the ability to gracefully recover from packet loss is far greater with the system requirements discussed here. Computational complexity is also described. Then, algorithm performance on both unquantized and quantized data is presented.

4.5.1 Implementation Aspects

Coefficients at the lowest decomposition level are packetized within each subband, so that if a low frequency coefficient is lost, the high frequency coefficients corresponding to the same spatial location are not lost along with it. Note that this packetization strategy still permits progressive transmission, and if packet are lost and reconstruction is performed, only one level of progression is lost. When packetized, data is assumed to be interleaved within subbands, so that large contiguous areas of loss within a subband are avoided, thereby allowing reconstruction using neighboring coefficients. Coded data is transmitted in a predetermined order, and sequence numbers as inserted into each packet, enabling the detection of lost packets. Thus the locations of lost coefficients are known at the decoder.

Any source coding techniques can be applied to the subband coefficients, provided that the resulting stream can be packetized such that loss of a packet does not affect decoding of subsequent packets. For example, scalar quantization, vector quantization, or transform coding of the subband coefficients followed by variable-length coding are acceptable providing only that the variable-length codes are self-contained within packets. Huffman coding of coefficients is acceptable when packet boundaries fall between codewords, while segmenting an arithmetically coded stream across several packets is not acceptable, as loss of any packet destroys synchronization at the decoder. Proposed subband coding techniques that can immediately be used with the described intraband coding, interleaving, and packetization requirements include [9]–[14].

The compression efficiency is effected by the packetization requirements described in the previous paragraph if fixed-length packets are used. Assume that a packet contains P data bits and that P exceeds the length of the longest codeword (packet overhead bits are not included in the calculations, because they are present regardless of how the data bits are placed in packets). With the average codeword length in coded subband i given by $\bar{l}_i$, assuming that all codeword lengths are equally likely to overflow a packet yields an expected fractional increase in the data stream of $\{(P - \lfloor \lfloor P/\bar{l}_i \rfloor \bar{l}_i \rfloor)/P\}$, where $\lfloor \cdot \rfloor$ is the floor operator and the numerator represents the average number of bits per packet that are unused when transmitting subband i. Note that if the codeword lengths are fixed, $\bar{l}_i = l_i$ and the increase can be zero if P is a multiple of all l_i's. For example, for the fixed-length technique presented in [9] (fixed-rate lattice vector quantization, for *lena* at 0.25 bits/pixel, PSNR = 31.4 dB), with ATM packets with $P = 384$, the increase is 14%. For the variable-length technique presented in [14] (entropy-constrained lattice vector quantization, for *lena* at 0.136 bits/pixel, PSNR = 30.9 dB), again using ATM packets the increase is 4%. The smaller increase in the cited variable-length technique is due to shorter average codeword lengths, caused by using vector dimensions smaller than those used in the fixed-length technique. With shorter average codeword lengths, the number of unused bits is on average smaller. The fixed increase in data rate with the proposed reconstruction technique does not limit the amount of lost data that can be reconstructed. In contrast, FEC or retransmission limit the amount of data that can be reconstructed to at most the amount of overhead information, and possibly less, depending on spacings of lost data in case of FEC and transmission success in the case of retransmission. As such, the proposed reconstruction technique is more efficient in dealing with higher loss percentages.

Computational overhead at the decoder is a function of the number of decompositions and the filter length. For a five tap symmetric filter, the decoder overhead per percentage of lost coefficients across all bands is 0.5% for one level of decomposition. For longer filters and higher numbers of decomposition levels, these values decrease.

4.5.2 Performance Simulations

Algorithm performance was evaluated by reconstructing random loss of three coefficient groupings across all subbands at all decomposition levels: single coefficients, vectors of length 4, and blocks of size 2 × 2. These three groupings may appear in source coding of the subbands using scalar quantization, vector quantization, and transform coding, for example.

The most accurate reconstruction is achieved when all coefficients used in interpolation and derivative estimation are known. If there is a horizontal or vertical vector of low frequency coefficients lost, near-optimal reconstruction is also achieved because half-derivatives can be used. If coefficients are coded in horizontal or vertical vectors, better reconstruction is obtained if vectors are staggered down rows or columns. In the case of length 4 vectors, the LH and HL bands were assumed vectorized in the high frequency direction to minimize the number of adjacent lost coefficients in the low frequency direction.

4.5.3 Unquantized Subband Performance

The *couple* image was subband decomposed to 1, 2, and 3 levels using both odd and even length quadrature mirror filters as given in [6, 7]. In general, the reconstructed LH and HL bands provide accurate information for edge placement. Low frequency reconstruction performs well on horizontal, vertical, and strong diagonal edges. To demonstrate algorithm performance on diagonal edges and to contrast the performance of the proposed algorithm with standard bicubic interpolation, an enlarged section of the curtains in the *couple* image is shown in Figure 4.8. In general, one-directional high frequency patterns are maintained, and spatial masking tends to reduce the visual effects of errors in multidirectional patterns. Setting lost HH coefficients to zero produces negligible visual effects. One- and two-level decompositions of a segment of *couple* suffering 10% random vector loss and reconstructed using the algorithm are shown in Figures 4.9 and 4.10.

(a)   (b)

(c)   (d)

Figure 4.8   Segment of the *couple* image demonstrating reconstruction of diagonal edges. (a) original, (b) synthesized segment with no reconstruction of coefficient loss in the low frequency band, PSNR=15.5dB (c) synthesized segment with low frequency co-efficients reconstructed using standard bicubic interpolation, PSNR=32.9dB (d) synthe-sized segment with low frequency coefficients reconstructed using edge-model based surface generation, PSNR=37.1dB

(a)

(b) (c)

Figure 4.9   Segments from unquantized *couple* (one decomposition level, 5-tap QMF) with 10% random vector loss in all subbands: (a) analyzed image with loss (b) synthesized image with no reconstruction, PSNR=16.2dB (c) reconstructed, PSNR=34.1dB

(a)

(b)   (c)

Figure 4.10 Segments from unquantized *couple* (two decomposition levels, 5-tap QMF) with 10% random vector loss in all subbands: (a) analyzed image with loss (b) synthesized image with no reconstruction, PSNR=16.4dB (c) reconstructed, PSNR=32.6dB

Loss and reconstruction of individual coefficients produces slightly higher PSNRs than vectors, which in turn have slightly higher PSNRs than 2×2 block loss. Randomly lost coefficients are most likely to have the highest number of known coefficients required in interpolation present. Randomly lost vectors are only missing coefficients in one direction, while blocks require the most coefficient estimation for use in interpolation. A plot of PSNR versus percentage loss for three types of loss for both one and two decomposition levels is shown in Figure 4.11.

The visual effects of errors in low frequency reconstruction change as the number of decomposition levels increases. At one decomposition level, errors in edge reconstruction are visible as small, sharp discontinuities in edges. At two and three levels, edge reconstruction errors are visible as slight or moderate blurring in the vicinity of an edge, caused by multiple levels of upsampling and filtering. Figure 4.11 indicates that there is a difference of approximately 2 dB between reconstructed quality of the same types of loss at one and two levels of decomposition on unquantized data.

4.5.4 Quantized Subband Performance

To evaluate performance on quantized as well as unquantized coefficients, the low frequency subband was quantized using scalar quantizers designed for a Gaussian distribution using the Lloyd-Max algorithm, and the high frequency subbands were quantized using scalar quantizers designed for a generalized Gaussian distribution with parameter 0.7, again using the Lloyd-Max algorithm. Bit allocations were chosen for constant quality, specifically, the PSNR of the synthesized quantized subbands was 35 dB. This value was selected because the corresponding bit allocations provided the highest compression ratio in which objectionable quantization artifacts (e.g., splotching and excessive graininess) were not visible in a one decomposition level synthesized image. (The data was quantized to evaluate algorithm performance, not to demonstrate a compression technique. In an actual implementation, variable length coding would follow the quantization to provide more realistic compression ratios. The bit rates quoted do not include entropy coding.)

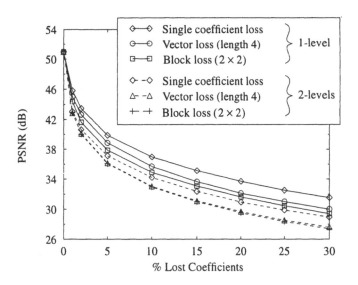

Figure 4.11   Reconstruction performance on unquantized *couple* for one and two decomposition levels and random loss of isolated coefficients, 4-vectors, and 2×2 blocks.

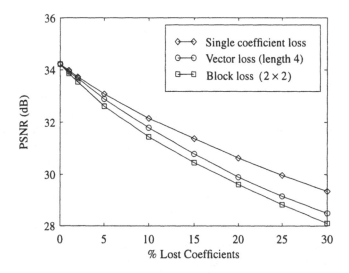

Figure 4.12   Reconstruction performance on quantized *couple*, using 5-tap filter with one decomposition level, for random loss of isolated coefficients, 4-vectors, and 2×2 blocks.

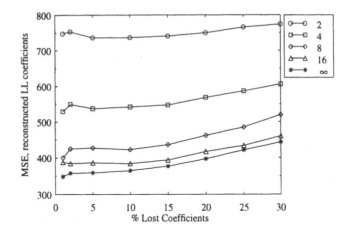

Figure 4.13   Mean-squared error of reconstructed LL coefficients (*couple*, 5-tap filter, one decomposition level) versus percentage lost coefficients for four quantizers applied to the high frequency bands. The lowest line represents the MSE of reconstructed coefficients using unquantized coefficients.

When the subband data is quantized, the PSNR spread drops, as illustrated in Figure 4.12 for one decomposition level. Quantization of the high frequency subbands at the lowest decomposition level affects the reconstruction quality of the LL subband. Quantization that is too coarse destroys the edge classification patterns used to identify edges. Figure 4.13 compares the mean squared error of the reconstructed LL coefficients at 1–30% random loss for four different high frequency bit allocations. The LL band bit allocation is constant at 5 bits/sample (32-level quantizer), while the HL and LH bands are quantized to 2, 4, 8, and 16 levels (the HH band is omitted). With 16 quantization levels for the high frequency bands, the MSE is within 20% of the non-quantized MSE. Visually, errors begin to appear in a 4-level quantizer, and a 2-level quantizer produces unacceptable reconstructed edges. Four levels is also the minimum required to avoid excessive graininess in the synthesized image, so it appears that provided the bit allocation does not produce graininess, the reconstruction algorithm will work. For two decomposition levels with 7 bits/sample in the LL band, reconstruction errors induced by high frequency band quantization appear at 4 bits/sample in the LH and HL bands at the lowest decomposition level, while graininess appears at 3 bits/sample. Reconstruction errors in two decomposition levels tend to be minimized in visual impact due to two stages of upsampling and low-pass filtering, compared to only one in a one level decomposition.

One- and two-level quantized decompositions of a segment of *couple* suffering 10% random vector loss and reconstructed using the algorithm are shown in Figures 4.14 and 4.15.

(a)   (b)

Figure 4.14   Segments from quantized reconstructed *couple* (one decomposition level, 5-tap QMF) with 10% random vector loss in all subbands (same loss as shown in Figure 4.5(a)): (a) 2.25 bits/sample, reconstructed PSNR=30.2dB (with no loss, PSNR=32.8dB) (b) 3.5 bits/sample, PSNR=32.6dB (with no loss, PSNR=38.0dB).

4.6 SUMMARY

Transmission of subband-coded images over imperfect networks can include error concealment strategies such as that described in this chapter to produce high quality images even when subband coefficients are lost in transmission. The described reconstruction algorithm for hierarchical subband-coded images uses interband relationships based on QMF properties to accurately reconstruct edge structures in the visually-important lowest frequency band, thus assuring minimal visual distractions in the synthesized reconstructed image. Surface fitting using bicubic interpolation provides the smooth

130

(a)   (b)

Figure 4.15   Segments from quantized reconstructed *couple* (two decomposition level, 5-tap QMF) with 10% random vector loss in all subbands (same loss as shown in Figure 4.6(a)): (a) 2.31 bits/sample, reconstructed PSNR=30.1dB (with no loss, PSNR=33.7dB) (b) 3.5 bits/sample, PSNR=31.5dB (with no loss, PSNR=37.6dB).

characteristics required of the lowest frequency band while providing a simple technique for accurate interpolation, by simply adapting the two-dimensional interpolation grid. The algorithm is not limited to QMFs and is easily adapted to work with any linear-phase filters.

Error concealment for subband-coded images is both computationally simple and provides reconstructed images of high visual quality. As such, subband coding can be used to provide spatial scalability and progressive transmission over channels which have not been traditionally considered for image transmission.

References

[1] Shacham, N. and McKenney, P. Packet recovery in high-speed networks using coding and buffer management. *Proceedings IEEE Infocom 90*, Los Alamitos,

CA, 1990, Vol. 1, pp. 124–31.

[2] MacDonald, N. Transmission of compressed video over radio links *BT Technology Journal*, Vol. 11, No. 2, April 1993, pp. 182–5.

[3] Wang, Y. and Ramamoorthy, V. Image reconstruction from partial subband images and its application in packet video transmission *Signal Processing: Image Communication*, Vol. 3, No. 2–3, pp. 197–229, June 1991.

[4] Johnsen, O., Shentov, O. V. and Mitra, S. K. A Technique for the Efficient Coding of the Upper Bands in Subband Coding of Images *Proc. ICASSP 90*, Vol. 4, pp. 2097–2100, April 1990.

[5] Mohsenian, N. and Nasrabadi, N. M. Edge-based Subband VQ Techniques for Images and Video *IEEE Trans. Circuits and Systems for Video Technology*, Vol. 4, No. 1, pp. 53–67, Feb. 1994.

[6] Woods, J. W. *Subband Image Coding* Kluwer Academic Publishers, Boston 1991.

[7] Johnston, J. D. A Filter Family Designed for Use in Quadrature Mirror Filter Banks *Proc. IEEE ICASSP 80*, vol. 1, pp. 291–4, Denver, CO, April 1980.

[8] Kay, S. M. *Fundamental of Statistical Signal Processing: Estimation Theory* Prentice Hall, Englewood Cliffs, New Jersey, 1993.

[9] Tsern, E. K. and Meng, T. H.-Y. Image Coding Using Pyramid Vector Quantization of Subband Coefficients *Proc. IEEE ICASSP 94*, vol. 5, pp. 601–4, Adelaide, Australia, April 1994.

[10] Hung, A. C. and Meng, T. H.-Y. Error Resilient Pyramid Vector Quantization for Image Compression *Proc. International Conference on Image Processing*, vol. 1, pp. 583–7, Austin, TX, November 1994.

[11] Ramchandram, K. and Vetterli, M. Best Wavelet Packet Bases in a Rate-Distortion Sense *IEEE Trans. Image Processing*, Vol. 2, No. 2, pp. 160–75, April 1993.

[12] Antonini, M., Barlaud, M., Mathieu, P. and Daubechies, I. Image Coding Using Wavelet Transform *IEEE Trans. Image Processing*, Vol. 1, No. 2, pp. 205–20, April 1992.

[13] Gharavi, H. and Tabatabai, A. Sub-band Coding of Monochrome and Color Images *IEEE Trans. Circuits and Systems*, Vol. 35, No. 2, pp. 207–14, February 1988.

[14] Senoo, T. and Girod, B. Vector Quantization for Entropy Coding of Subbands *IEEE Trans. Image Processing*, Vol. 1, No. 4, pp. 526–33, October 1992.

5 VIDEO CODING STANDARDS: ERROR RESILIENCE AND CONCEALMENT

Mark R. Banham and James C. Brailean

Motorola
Chicago Corporate Research Laboratories
1301 E. Algonquin Rd.
Schaumburg, IL 60196-1078
{banham, brailean}@ccrl.mot.com

Abstract: The emergence of digital video compression as a means to enable visual communications has been driven largely by the efforts of voluntary international standards organizations. The standards developed by these organizations have substantially increased the use of video in many different applications. However, transmitting standardized video bitstreams in error prone environments still presents a particular challenge for multimedia system designers. This is mainly due to the high sensitivity of these bitstreams to channel errors. This chapter examines the existing standards for video compression and communication, and details the tools for providing error resilience and concealment within the scope and syntax of these standards.

5.1   INTRODUCTION

The importance of standards cannot be overstated in terms of their effect on driving technology in the marketplace. This is particularly true of standards that permit the use of communication tools created by different manufacturers to interact seamlessly. The standards that address the novel field of visual communications fall under two different voluntary standards organizations. The first of these is the International Telecommunications Union/ Telecommunications Standardization Sector (ITU-T) which is a specialized agency of the United Nations. The second is the International Organization for Standardization/International Electrotechnical Commission (ISO/IEC), a worldwide federation of national standards bodies. Each of these organizations has been responsible for important standards and recommendations for video coding and transmission. Table 5.1 provides an overview of the video coding standards discussed in this chapter. These standards make up the majority of video codec solutions in use by industry today. There are a number of associated system and multiplex standards, which are also discussed in this chapter, as it is often the system layer which ultimately determines the performance of video communication in the presence of errors.

STANARDS ORGANIZATION	STANDARD NAME	YEAR OF ADOPTION	FUNCTIONALITY DESCRIPTION
ITU-T	H.261	1990 (revised 1993)	Videoconferencing 128-384 Kbps
	H.263	1996 (revised 1998)	Videoconferencing 10-384 Kbps
ISO/IEC	MPEG-1	1992	Digital Storage Media 1-2 Mbps
	MPEG-2	1994	Broadcast 4-6 Mbps
	MPEG-4	1999	Content-Based Interactivity 10 Kbps - 4 Mbps

Table 5.1: Summary of Video Coding Standards.

5.1.1   Applications of Video Coding Standards in Error Prone Environments

Transmitting standardized video in the presence of errors is not a novel concept by any means. For decades, analog NTSC and PAL video has been modulated and transmitted over noisy broadcast channels, often with significant degradation to video quality. For analog broadcast applications, however, the video quality typically degrades gracefully as a function of the receiver's distance from the transmitter. In the world of digital video, an entirely different situation arises. Compressed digital video is more susceptible to the effects of channel noise, when bit errors are not entirely removed by error correction. Because information is typically predictively coded, errors can propagate through a decoded sequence over time, making it difficult to hide the errors from the viewer.

The digital video compression standards discussed in this chapter provide the tools to send a much larger amount of video data, using much less bandwidth, than the old analog methods. This alone warrants the use of digital video compression for a variety of applications, some of which require transmission over noisy channels.

Perhaps the most prevalent requirement of video compression standards in error prone environments is the ability to handle mobile communications. With the development cellular systems supporting higher capacity digital data, videophone applications which require resilient standards are likely to emerge. Videoconferencing has already become very prevalent over low bandwidth packet-switched networks, which are subject to a different type of errors than those of wireless channels. Video data can be subjected to noise in other applications as well. Some of these include the acquisition of video from noisy digital storage media and the broadcasting of digital video over satellites or other delivery systems.

Given an inherent resilience to errors, the best standards can provide excellent quality at the decoder even in the presence of a certain level of errors. Table 5.2 provides a listing of the most prevalent applications discussed within the standards bodies that will require error resilient video standards.

Domain	Application
Mobile Video Communications	Personal Video Communications, Dispatch Video Services, Group Video Calls
Mobile Video Surveillance	Law Enforcement, Home or Office Security, In-Home Monitoring
Mobile Command and Control	Police, Fire, Rescue, Ambulance: Remote to Central Station Communication
Digital Storage Media	Digital Versatile Disk, Computer Hard Disk Storage and Playback
Internet Video	Streaming Video, IP Multicasting, PC Desktop Videoconferencing
Broadcasting	HDTV, Digital Broadcast Satellite

Table 5.2: Applications Using Robust Video Coding Standards.

5.2 STANDARDIZED VIDEO COMPRESSION AND TRANSMISSION

The video standards discussed in the chapter are all quite closely related in terms of their fundamental technology. In fact, all may be classified as hybrid Discrete Cosine Transform (DCT)/motion-compensated coding standards. The general approach to compression taken by all of these standards is illustrated in Figure 5.1.

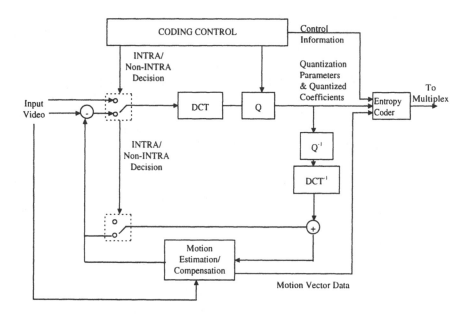

Figure 5.1: Standard Video Source Encoder Diagram.

The elements above, including the DCT and Motion Estimation and Compensation, are essential parts of each of the standards discussed here. This approach has proven to be the most efficient and straightforward method for video compression. However, some limitations do make it difficult to communicate this information in the presence of errors. These limitations can be overcome with some tools and algorithms enabled by the standards themselves.

5.2.1   Compression Technology and Its Limitations

The need to achieve high compression ratios in order to transmit digital video results in bitstreams that are highly sensitive to errors. This is a consequence of the relative information content of each bit increasing with the compression ratio.   One requirement of wireless video transmission for personal communications would be that the compression ratio applied to the video exceed 5000:1, when compressing source material obtained at ITU-T Rec. 601 resolution. Therefore, one bit error could have the impact that 5000 bit errors would have on the uncompressed data. While the relationship between errors in the compressed and uncompressed domains is non-linear, this example illustrates, in general terms, the sensitivity that these standards can have to errors.

Predictive coding and error propagation. The essential approaches used to reduce the entropy of the video source for compression are reduction of redundant information and removal of irrelevant information. Because most video sequences are highly correlated in both space and time, it is possible to remove redundant information by using predictive coding. This is of particular concern when studying the impact of errors on digitally compressed video. When an error is encountered in the compressed bitstream, it is very likely that the effects of that error will propagate to neighboring spatial blocks and frames due to predictive coding. Techniques provided by some of the standards discussed in this chapter are specifically targeted at eliminating, or at least *containing* the propagation of errors to a small region of time or space.

Motion Compensation. Motion compensation is the single most important method used by all video compression standards to reduce the overall bitrate. In each of the standards, the details are different, but all use block-based motion estimation and compensation to predict the picture content from frame to frame. Motion vectors are coded and accompanied by DCT encoded prediction error blocks. The propagation of motion-compensated predictive information only stops when an INTRA, or I, coded macroblock is transmitted. Such macroblocks do not contain any prediction from prior frames, but only code the spatial information at the current time instant.

Motion Vector and Quantization Parameter Coding. The motion vectors themselves need to be compressed efficiently in order to achieve low bitrates. This is typically done by predicting the value of motion vectors as a function of the neighboring macroblocks' motion vectors. Because the motion vectors in a scene are often correlated across the frame, they can easily be represented as a difference from motion vectors that have already been coded. A small motion vector prediction error then needs to be transmitted as well. In a similar manner, the quantization parameters used for coding the prediction error and I macroblock information is also correlated across the frame, and is often predicted from prior macroblocks. Errors in predictively coded motion vector and quantization parameter data will propagate until the prediction is reset using unique start codes in the bitstream.

Variable length codes and resynchronization. The entropy codes representing various syntax elements, like motion vector data, quantized DCT coefficients, and control information have been optimized to provide excellent compression in all of the standards. The fact that the codewords used almost all come from variable length code (VLC) tables, means that the boundaries between codewords are implicit in the decoder. A variable length codeword decoder reads bits until a full codeword is encountered, then it translates that codeword into a meaningful symbol, and begins decoding a new word. When there is an error in a variable length codeword, the decoder may not be able to detect that error, but will rather decode an incorrect symbol. In cases where the VLC tree is almost full, it is very easy to emulate valid VLCs in the presence of bit errors. These errors may never be detected until a unique resynchronization point or start code is encountered in the bitstream.

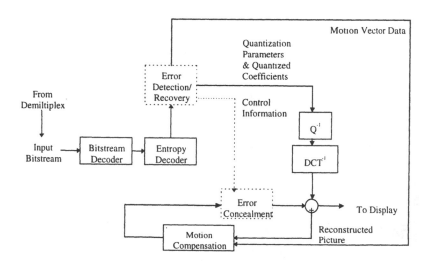

Figure 5.2: Standard Error Resilient Video Decoder Diagram.

As seen in all the limitations discussed here, resynchronization is a very important element when operating in the presence of errors. A decoder must be able to detect, recover from, and conceal errors if it will operate in error-prone environments. Figure 5.2 shows a standard error resilient video decoder including these functions. The different approaches to resynchronization within the video bitstream and recovery of lost data will be explored later in detail for each of the relevant compression standards.

140

5.2.2 Multiplexing Audio, Video, and Control Information

In order to communicate synchronized audio and video information, it is necessary to group the output of the audio and video encoders into meaningful packets, and multiplex those packets over a channel. These packets of audio and video information are generally accompanied by control information indicating certain parameters about the operation of the audio and video codecs, and possibly accompanied by independent data from other applications. Error correction is often applied at the multiplex layer. The output bitstream from the multiplexer is the one that is actually exposed to channel errors.

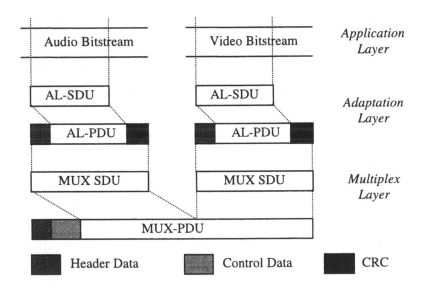

Figure 5.3: General Structure of System Multiplexer.

The general structure of a system multiplexer is shown in Figure 5.3. The original bitstreams from an audio and video encoder are partitioned into packets at the Adaptation Layer (AL). These packets are called Service Data Units (SDUs), and when headers and error control are added, they become Protocol Data Units (PDUs) which may be exchanged between peer layers of the protocol at the transmitter and receiver.

Variable vs. fixed multiplex packet structures. The MUX PDUs seen in Figure 5.3 may be of fixed or variable length. Some standards use a fixed length approach, resulting in good synchronization between transmitter and receiver. This requires the expenditure of extra stuffing overhead, however, when the application layer does not have data ready for transport. Variable length packets are used by other standards, and these can utilize control and overhead bits more efficiently, but also require the transmission of special flags to indicate the boundaries of MUX-PDUs to the demultiplexer. The trade-offs between fixed and variable multiplex strategies have been examined within the standards bodies, and it is generally accepted that a variable size packet structure permits much better coding efficiency and flexibility. The drawbacks of synchronization flags can also be accommodated by new robust approaches, which will be discussed later.

Protecting the multiplex layer. The error resilience of a multiplex layer is crucial if proper communication is to be maintained. If an error appears in the header of a MUX-PDU, it is possible that the demultiplexer could misinterpret the location or content of the AL-PDUs contained within that MUX PDU. This could result in incorrect data being passed through the AL to the Application Layer. An extreme example of this would be sending audio data to the video decoder. Clearly, this would result in significant degradation to the video quality, which would be unacceptable for use in a real system.

Therefore, it becomes very important to protect the multiplex header from corruption by errors. In the presence of severe bursts or packet losses, it must also be possible to detect that a MUX header is corrupted so that the entire MUX-PDU may be discarded. This is almost always preferable to passing data from the wrong application layer to the video decoder.

Errors in the adaptation layer. It is often the case that errors in the channel will hit data within the payload of the MUX-PDU, but not the MUX header itself. In these cases, it is still possible to determine which bits apply to the video AL-PDU, and to pass them to the Adaptation Layer. The AL-PDU often contains some sort of cyclical redundancy check (CRC) code that can be used to detect the presence of errors.

Errors in the AL may be treated in one of two ways. The AL-SDU may be passed to the video decoder, or it may be discarded. In either case, it is necessary to have a video decoder that is equipped to detect missing or corrupted data, and to conceal it appropriately. There is no fixed formula for how to treat corrupted video data at the Adaptation Layer, but it is certainly beneficial to take advantage of error detection capabilities supplied by the system before deciding how to decode packets of video information.

5.2.3 Error Prone Transport Systems for Video

Today, the major distribution networks for video include the Public Switch Telephone Network (PSTN), Integrated Services Digital Network (ISDN), broadcast, and the Internet. Currently, videoconferencing is primarily conducted utilizing either PTSN or ISDN. Typically, these two networks display a very low probability of bit errors. Similarly, a broadcast system is designed to provide essentially error-free delivery of video within a certain service area. The Internet, however, is a network where information can be lost during transmission. Currently, the majority of Internet applications which utilize video are database retrieval or streaming applications.

Wireless video communication over land mobile and cellular systems is only beginning to emerge. First generation cellular systems are based on analog technology. Delivery of digital data is limited to a maximum of 14.4 Kbps under extremely favorable channel conditions. A substantial delay is also associated with this 14.4 Kbps throughput. Second Generation, or 2-G cellular systems such as the Global System for Mobile communications (GSM) in Europe, or Code Division Multiple Access (CDMA) in the US and Japan, are beginning to provide higher data rates using digital technology. However, these systems are still primarily designed for voice data. Therefore, the higher data rates are usually accompanied with long delays that substantially reduce the usefulness of the video.

One group of wireless systems that are beginning to deliver useable video includes the cordless telephone systems in Japan and Europe. The Personal Handiphone System (PHS) of Japan is a four slot Time Division Multiple Access (TDMA) system that provides 32 Kbps/slot. Several portable video concepts that operate over this system have been introduced. A similar TDMA system called Digital European Cordless Telephone (DECT) exits in Europe. In both cases, the user is provided with a 32 Kbps unprotected channel. In these systems, the range of the portable device is confined to an area that is near the transmitter.

In the future, desktop videoconferencing and TV programming will also utilize the Internet. However, these bandwidth intensive applications may cause more degradation to reliable transmission over the Internet. As new Third Generation, or 3-G, wideband cellular standards become available, applications such as mobile videoconferencing for business will likely emerge. Since the bandwidth available using these new wireless standards will much greater, the quality of the video be greatly improved. However, the error conditions of these new systems will still be extremely severe. Thus, error prone transport systems will be in wide use for the foreseeable future.

Error characteristics of mobile channels. When describing a communication channel, an additive white Gaussian noise (AWGN) model is often assumed. The AWGN model usually provides a reasonable starting point to analyze basic performance relationships.   But, for most practical channels, where signal propagation takes place near the ground, the AWGN, or free space, model is inadequate for describing the error characteristics of the channel.

In a wireless mobile communication system, a signal can travel from the transmitter to the receiver over multiple reflective paths. This phenomenon is referred to as "multipath" propagation. The effect can cause random fluctuations in the received signal's amplitude and phase, which is referred to as multipath fading. In 2-G cellular systems, the effects of multipath fading can typically result in as much as a 12-15 decibel (dB) drop in the received signal-to-noise ratio (SNR).   Since SNR is directly related to the bit error rate (BER) of the receiver, multipath fading results in large variations in the BER performance of the channel.   In other words, the error properties of a mobile channel are characterized as "bursty".   The frequency and length (generally measured in time) of the bursts are dependent upon the type of terrain (i.e., urban, mountainous, etc.) and the speed of the mobile unit [1].

Error characteristics of wired channels. The error characteristics of a circuit-switched wireline POTS, or "plain old telephone service", transmission are around 10^{-6} random BER in the worst cases. This applies to communication rates near 28.8 Kbps, which is supported by the V.34 modem standard. This level of errors is quite weak, and generally will only cause moderate disruption to a video bitstream. The modem protocols used to communicate over POTS lines also provide further error protection, dropping the error rate to near zero in most cases.

When communicating over wired packet-switched networks, a different type of errors can result. These errors are packet losses resulting from network congestion, and are a function of the transport layer protocols being used. Some transport protocols, such as TCP, or Transmission Control Protocol, provide a reliable end-to-end connection, and will use retransmission to guarantee delivery of data. However, this results in heavy delay when the network is congested. UDP, or User Datagram Protocol, uses no retransmission, and thus provides an unreliable, connectionless service. When the datagrams that make up a message are sent, there is no verification that the have reached their destination. So, when a predetermined period of time elapses, an erroneous packet will be lost. It is possible to have a widely varying level of errors in the bitstream when using UDP over the Internet or a LAN.

5.3 ITU-T SERIES H RECOMMENDATIONS - AUDIOVISUAL AND MULTIMEDIA SYSTEMS

The ITU-T has dedicated a large portion of its standards activities to the development of recommendations for visual communications. These recommendations fall under the category of Series H Recommendations, which deal primarily with the line transmission of non-telephone signals. The Series H Recommendations provide protocols for a variety of issues, including source coding, multiplexing, and channel coding and decoding. There are three Series H system standards that address videoconferencing: H.320, H.324, and H.323.

H.320 and H.324 deal with circuit-switched communications, and are representative of the typical system characteristics of a videoconferencing call. H.323 uses the same source codecs as H.324, but provides a multiplexing protocol for packet-switched communication. Packet loss errors can create problems for video decoders in the same way that burst errors can create problems when using a circuit switched standard over a wireless link. The methods discussed later for the treatment of burst errors in the video layers of H.320 and H.324 are directly applicable to the use of H.323. However, discussion of system layers here is limited to the circuit-switched standards.

5.3.1 *H.320 - Narrow-band Visual Telephone Systems and Terminal Equipment*

H.320 was the first overall system standard for videoconferencing [2]. It specifies the use of a number of other standards for source and control coding and multiplexing, including, but not limited to, Recommendations H.261 and H.263 for video coding, G.723 for audio coding, H.245 for control, and H.221 for multiplexing. H.320 is probably the most widely deployed system standard for wireline videoconferencing today. It was first decided by the ITU-T in 1990, and has undergone several revisions since then. This is the protocol used for specifying interoperability of multimedia terminals communicating at ISDN rates.

H.221 - Frame structure for a 64 to 1920 Kbps channel in audiovisual teleservices. This recommendation defines a multiplexing structure for single or multiple ISDN channels with a constant framing structure [3]. It is dedicated to joining audio, video and control information into fixed size packets which can then be transmitted using variety of user data rates from 300 bits/s up to almost 2 Mbps.

The advantage of H.221 is that it is relatively simple and can be easily implemented on most microprocessors or dedicated hardware systems. The fixed size MUX-PDU structure allows for simple synchronization between the transmitter and the receiver. The configuration of the multiplex can be changed at 20ms intervals, using double-error-correcting codes for the control information. In networks where it is not possible to obtain octet level synchronization from external control data, H.221 can be used to derive this level of synchronization. So, it is possible to recover the location of multiplex frame boundaries if synchronization is lost due to errors or bit slippage.

H.261 - Video codec for audiovisual services at p x 64 Kbps. The earliest standardized video codec for videoconferencing was H.261 [4]. This was decided by the ITU-T in 1990, and was revised in 1993. H.261 is a basic hybrid DCT motion-compensated video codec. It uses many of the same video coding tools as the other standards discussed here. H.261 does not support half-pel motion estimation, but instead provides a loop filter for improved picture quality. This typically does not produce comparable quality to H.263, but is less computationally expensive. It also does not support any of the advanced coding tools in the annexes of H.263, which will be discussed later.

H.261 does provide a very rudimentary style of error resilience, which is in the form of a (511,493) BCH code applied to bit frames of video data. This forward error correction (FEC) can be used to protect the video bitstream against moderate random errors that may appear on a wireline call. The BCH framing provided by H.261 is seen Figure 5.4. It can be seen in this figure that the frame alignment bits, S_i, form a specific pattern over every eight 512-bit frame intervals. These bits can be used to verify synchronization within the video. The 18 parity bits can be used to detect up to three and correct up to two errors in the 493 preceding bits (this includes the video data or fill bits and the fill indicator, or Fi, bit). The Fi bit is used to indicate the fill mode for use in stuffing ones in the bitstream when the bits in the video buffer are temporarily exhausted.

146

S_i = Frame alignment bit.

$(S_1S_2S_3S_4S_5S_6S_7S_8) = (00011011)$

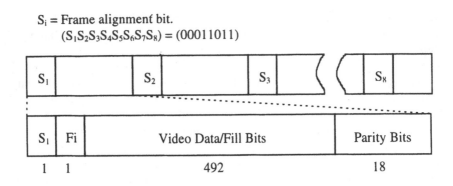

Figure 5.4: Error Correction Framing in H.261.

One important tool for decoder error resilience is the location of resynchronization points in the bitstream. These points can be used to localize the errors and resume predictive decoding. H.261 has a unique Group of Blocks (GOB) structure. Each GOB has a header containing a unique code word that may be found in the bitstream by a simple correlator. The GOB structure indicates the minimum regions within which errors may be contained.

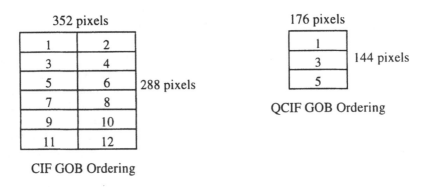

Figure 5.5: GOB Structure in H.261.

Figure 5.5 shows the GOB structure for common intermediate format (CIF) and quarter CIF (QCIF) frames using H.261. The dimensions of the luminance, or Y channel are shown in terms of numbers of pixels. The two accompanying chrominance channels for these formats (Cr and Cb) are subsampled to half the size of the Y channel in each dimension. Each GOB consists of 33 spatially contiguous macroblocks. Thus, the smallest region into which an error in an H.261 video bitstream can be localized is 33 macroblocks. While the tools mentioned here provide some amount of error resilience, H.261 is not considered to be a standard which is robust to errors, and it is not recommended for use over wireless or other error-prone channels.

5.3.2   H.324 - Terminal for Low Bit Rate Multimedia Communication

H.324 is a more recent system level standard for videoconferencing developed by the ITU-T [5]. It addresses the communication of video, audio, and data over V.34 modem connections using an analog POTS telephone line. The advent of improved compression technology made it possible to communicate at the very low bitrates supported over POTS. Like H.320, H.324 defines the interoperability requirements of terminals used to conduct a multimedia call using a suite of supported standards. These standards include Recommendations H.261 and H.263 for video coding, G.723 for audio coding, H.245 for control, and H.223 for multiplexing.

H.324/Annex C - Multimedia telephone terminals over error prone channels. This annex is the result of efforts within the Mobile Group of ITU-T SG16, which worked to produce a set of revisions dedicated to improving the system level transmission of videoconferencing calls [6]. The main benefits provided by Annex C of H.324 are:

- The mandatory use of NSRP (Numbered Simple Retransmission Protocol) which improves the transfer of control information in error prone environments by adding a sequence number to SRP response frames, thereby eliminating possible ambiguity about which frame is being acknowledged.

- The use of robust versions of the multiplexer (H.223/Annexes A, B, and C) which help to guarantee the delivery of video and audio information to the correct decoders at different levels of robustness.

- Robust level setup procedure and procedure for dynamically changing levels of the multiplex during a call, allowing an adaptive use of error resilience overhead depending on channel conditions.

H.223 - Multiplexing protocol for low bit rate multimedia communication.
H.223 is different from its predecessor, H.221, in that it permits the transmission
of variable sized MUX-PDUs [7]. This requires the boundaries of MUX-PDUs
to be delineated by special flags. These flags are called High-level Data Link
Control (HDLC) flags. They are eight bits long, "01111110", and not unique to
the bits passed up from the Adaptation Layer. In order to make these flags
detectable, the transmitter must examine the MUX-PDU content before placing
HDLC flags on the boundaries of the packet, and stuff a "0" after all sequences
of five contiguous "1" bits. The receiver must then strip these stuffing or
"transparency" bits after finding the HDLC boundary flags.

It is easy to observe that the transparency method is not very robust to errors.
If an error appears on a stuffing bit, the decoder will not remove it, and it may
end up being passed to the video decoder. It is also possible that an error in an
HDLC flag would result in the failure to detect a MUX-PDU correctly, resulting
in either improper data being passed to the AL, or loss of the entire MUX
packet.

The robust annexes to H.223 are designed to provide improvement to this
structure in ways that are more resilient to errors. This is accomplished through
three new levels of robust operation, indicated by Annexes, A, B, and C. The
main benefit of communicating with these annexes is the better protection of
MUX-PDU headers as well as AL-PDU content.

H.223/Annex A – Communicating over low error-prone channels. Annex A is
the lowest level of robust communication, referred to as Level 1, just above
normal H.223, or Level 0, in its robustness [8]. The main improvement provided
by Annex A is the replacement of the 8 bit HDLC flags with a non-unique 16 bit
pseudo-noise synchronization flag. This flag may be detected in the decoder by
correlating the incoming bitstream with the known flag bits. It is possible to
have emulations of this flag in a normal bitstream, but the likelihood of this is
very small [9], and far outweighed by the improvement in communication gained
over error-prone channels [10]. This improvement comes from the elimination of
the transparency stuffing of Level 0, which led to such poor performance in the
presence of errors.

H.223/Annex B – Communicating over moderate error-prone channels. The
next stage in improving the operation of the multiplex is found in Annex B [11].
This annex, pertaining to Level 2 operation, increases the protection of the
MUX-PDU header itself. This header contains a multiplex code (MC) which
tells the receiver the exact contents of the mux packet. Errors in the MC can
cause the wrong data to be sent to the different Adaptation Layers, resulting in

significantly degraded performance, as previously discussed. The syntax of the Annex B MUX-PDU header is seen in Figure 5.6.

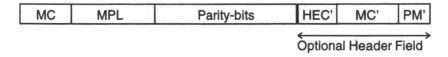

Figure 5.6: H.223/Annex B MUX-PDU Header Syntax.

The Annex B header uses a Multiplex Payload Length (MPL) field to provide an additional pointer to the end of a MUX-PDU over the Level 1 synchronization flags. A (24,12,8) Extended Golay Code then protects the MC and MPL fields. An optional Header Error Control (HEC') and repeated MC of the previous MUX-PDU (MC') and repeated Packet Marker of the previous MUX-PDU (PM') may also be included in this header. All of these components provide a more resilient means to maintain synchronization of the MUX-PDU boundaries and AL-PDU boundaries within.

H.223/Annex C – Communicating over highly error-prone channels. Building upon the previous two Levels, Annex C is used to provide Level 3 robustness of the multiplex [12]. This is accomplished by addressing protection of the AL-PDUs themselves. Annex C provides the following extensions to help protect the data in the AL-PDUs: error detection and correction, sequence numbering, automatic repeat request, and ARQ Type I and II retransmission capabilities.

These features may be applied to control, audio, and video logical channels, corresponding to the robust, or mobile Adaptation Layers: AL1M, AL2M, and AL3M. Annex C provides error detection using several different CRC polynomials, and Rate Compatible Punctured Convolutional (RCPC) Codes [13] for error correction. Optional interleaving is also supported along with retransmission. Given all of these enhancements, it is possible to provide further protection of the actual video data in the AL-PDUs. All of this comes at a substantial added complexity cost to that of lower levels of the protocol.

H.263 - Video coding for low bit rate communication. H.263 is the most popular video coding standard for low bitrate video communications in use today. It was first decided by the ITU-T in 1996, and was revised in 1998 [14]. H.263 provides all the functionality of H.261, with numerous additional benefits. The fundamental improvements over H.261 include the use of half-pel motion estimation and compensation, and overlapped block motion compensation. These two added features permit much more efficient coding of video, and allow the compression of QCIF resolution video sequences to rates as low as 10 Kbps, depending on scene content.

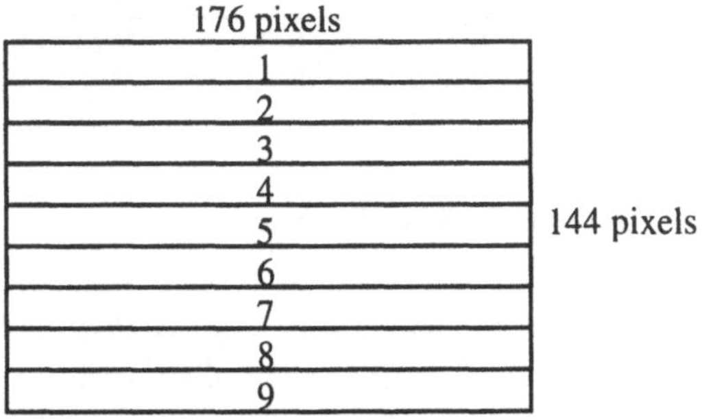

176 pixels

144 pixels

QCIF GOB Ordering

Figure 5.7: QCIF GOB Structure in H.263.

H.263 also provides a limited amount of error resilience, in the same manner as H.261, using the group of blocks structure and an optional (511,493) BCH code (H.263/Annex H). In H.263, GOB headers may only appear along the left edge of the picture. There are no data dependencies across GOBs, so they may be decoded independently of one another. The number of rows of macroblocks in a GOB is a function of the overall picture size. For CIF and QCIF, there is 1 row of macroblocks per GOB. For example, the GOB ordering for a QCIF resolution image in H.263 is seen in Figure 5.7. The smallest region that can be concealed in the presence of errors using only GOB resynchronization is thus one row of macroblocks (11 for QCIF, 22 for CIF).

There are a number of annexes to H.263 which are either specifically or tangentially related to improving error resilience when transmitting or storing video in error prone environments. These annexes are discussed next.

H.263/Annex K - Slice Structured Mode. As has been discussed, one of the principle means for providing error containment is found in the concept of groups of blocks, or slices. The ability to adaptively determine the location of synchronization points, and thus the slice structure, can allow for efficient use of bandwidth with good localization of important information in a video frame.

The Slice Structured mode of H.263 provides a fundamentally different slice structuring approach than that provided by the normal GOB structure of the H.263 standard. With this mode, it is possible to insert resynchronization markers at the beginning of any macroblock in a video frame. A slice is typically made up of the macroblocks appearing in raster scan order between slice headers. Therefore, it is possible to spend the overhead bits required for robust resynchronization in a more efficient way. Slice Structured mode start codes are accompanied by headers which contain all of the information necessary to reset predictively coded information and continue decoding. As with GOBs, no data dependencies in the current frame exist across slice boundaries, so they may be decoded independently of one another. A subset of the Annex K slice header is seen in Figure 5.8. In this header, the important fields for resynchronization are the SSC (Slide Start Code), MBA (Macroblock Address), and SQUANT (Quantizer Information). The SSC is a 17 bit unique codeword that can be found by running a correlator over the incoming video bitstream. The MBA can then be used to identify which macroblock appears next in the bitstream, and SQUANT resets the quantization step size, which is normally differentially encoded from macroblock to macroblock.

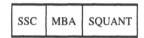

Figure 5.8: Syntax of Part of the H.263 Annex K Slice Header.

When using slice structured resynchronization, it is possible to send more resynchronization flags in the areas of the picture where there is a large amount of motion or spatial activity requiring many bits, and fewer where there is little motion or spatial activity. In this way, the expense of resynchronization overhead can be spread out in an even fashion over the bitstream. In the limit, the smallest region that can be localized with this approach is one macroblock. This minimum region of containment can be used to provide much better error resilience than the minimum provided by H.261 and normal H.263.

As an example, Figure 5.9 shows the location of GOB headers and Annex K slice headers in a P frame of QCIF video sequence coded at 48 Kbps with the relative spatial positions in which they appear. In this example, GOB headers are shown in white at every macroblock along the left edge of the picture. Annex K slice headers are shown in gray at every macroblock position after 512 bits have been spent in the bitstream. Figure 5.10 shows the relative positions of these headers in the bitstream portion corresponding to this frame. It can be seen that the GOB headers are unevenly distributed over the bitstream, while the slice headers appear at uniform locations. In the presence of a short burst of errors in the center of the image, the GOB method would need to discard a whole row of macroblocks, while the Slice Structured mode could often contain the errors to a few macroblocks.

152

Figure 5.9: Comparison of GOB Header Resynchronization (white) and Variable Resynchronization Markers (gray).

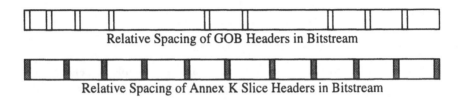

Figure 5.10: Bitstream Position of GOB headers vs. Annex K slice headers.

The Slice Structured mode also contains two submodes which can be used to provide additional functionality. The first is called the Rectangular Slice submode. This mode allows a slice to occupy a rectangular region of a specified width. It can be used to create different sized viewports in the image that can be coded independently of all neighboring slices. The second submode is the Arbitrary Slice Ordering submode, which allows slices to appear in any order in the bitstream. This can be used with different packetization schemes which allow for special priority coding of certain packets that might result in the delayed arrival of video slices at the decoder.

H.263/Annex N - Reference Picture Selection. Annex N of H.263 is the Reference Picture Selection mode, which allows a decoder to signal the occurrence of an error to the encoder through a back-channel, either a separate logical channel or a backward encoded video data multiplexer, if one exists. Given knowledge of an error at the decoder, the encoder can then be instructed to change the frame used for prediction at the next coded frame interval. The decoder stays in sync with the encoder through a series of optional acknowledgment signals sent through the backchannel.

The decoder requires an additional amount of picture memory in order to switch its predicted frame. This can be signaled to the encoder at the start of a call. The major benefit of this annex is that it provides the ability to conceal errors that would normally propagate in time until a series of INTRA macroblocks was coded. Given this quick concealment signaling, the overall quality of a video call in an error prone environment can be improved at the cost of some added memory and control complexity.

H.263/Annex R - Independent Segment Decoding. This mode provides a mechanism to decode a picture with no dependencies across slice or GOB boundaries having non-empty GOB headers. The benefit of this is that corrupted data cannot propagate across the spatial boundaries of predetermined segments in the frame. In Independent Segment Decoding mode, the motion compensation vectors may not point to macroblocks in segments of the reference picture that are different from the current segment. This provides an increased level of localization, helping to contain errors better. A major side effect of good compression is the high level of data dependencies due to predictive coding. So, this mode sacrifices compression efficiency by eliminating some of these dependencies for the benefit of better performance over error-prone channels.

There are several restrictions on the operation of Annex R. The first restriction is on the segment shapes themselves. Independent Segment Decoding mode cannot be used with the Slice Structured mode (Annex K) unless the Rectangular Slice submode of Annex K is active. This prevents the need for treating special cases when slices have awkward or non-convex shapes. The second restriction is that the segmentation, or shape of the slices or GOBs, cannot change from frame to frame, unless an INTRA frame is being coded. This constraint guarantees that the data dependencies can be limited to known regions, and that the motion data along boundaries will be treated correctly for each segment in the decoder.

H.263/Annex O – Scalability Mode. Scalability provides an inherent mechanism for error resilience in video bitstreams, although it is often present for providing progressive transmission properties. Annex O of H.263 provides a syntax for new picture types which are used for temporal scalability, spatial scalability, and SNR scalability. These types of pictures are used to generate enhancement layers, which can be decoded independently of one another. It is with enhancement layers that error resilience can be provided with this annex.

154

Data partitioning is a method that permits different treatment of certain classes of data depending on their relative importance to the source. When video data is separated into a base layer and enhancement layers, a natural class structure develops, and the data is automatically partitioned in the bitstream in terms of its relative importance. The base layer, which is always decoded, is the most important information, and thus can be protected better than the enhancement layers. The enhancement frames are less important, since future frames need not depend on these frames. So, if errors are detected in an enhancement layer, it is possible to continue decoding the base layer, and not display the enhanced data until it has had time to recover from the errors. The use of scalability with unequal error protection is not explicitly supported by H.263, but the concept may be implemented within systems having protocols which allow communication at multiple qualities of service.

5.4 ISO/IEC MPEG STANDARDS

The ISO/IEC MPEG standardization effort was originally started in 1988 to address the problem of compression of multimedia information. Since that time the MPEG committee has released the MPEG-1, MPEG-2 standards and will release its latest standard, MPEG-4, in January of 1999. Although the basic video compression techniques found in these three standards are the same, the applications they address are fundamentally different.

MPEG-1 addresses issues related to the storage and retrieval of multimedia information on a CD-ROM [15]. MPEG-2, which followed closely behind MPEG-1, is focused on high quality multimedia compression for use in broadcast applications [16]. The most notable uses of MPEG-2 include Direct Broadcast Satellite (DBS), Digital Versatile Disk (DVD) and High Definition Television (HDTV). It should be noted that the name MPEG-3 was reserved for HDTV. However, since the Grand Alliance found MPEG-2 to be useable for HDTV, it was decided to include HDTV as separate profile or option within MPEG-2. MPEG-4, which began as a low bitrate standard, was transformed into the first standard that truly addresses multimedia [17]. Text, graphics, real and synthetic video and audio are all sources which can be independently compressed and manipulated utilizing the object oriented syntax of this latest MPEG effort.

Error resilience and concealment work in MPEG has evolved in parallel to the compression activities. This is mainly a result of the applications addressed by each of the MPEG standards. For reasons discussed later in this section, very little attention was given to error resilience and concealment during the development of MPEG-1. After it was realized that MPEG-2 would be utilized over packet-based networks that would most likely be subjected to packet losses, work on error resilience and concealment was started within MPEG. From this early work in MPEG-2, error resilience and concealment have become major efforts within MPEG. The ability to access any MPEG-4 bitstream and decode it in real-time over an error prone network (e.g., wireless or packet-based networks) is a testament to this effort. The remainder of this section will be devoted to the MPEG standards and their error resilience and concealment capabilities.

5.4.1 MPEG Basics

The general block diagrams of a basic hybrid DPCM/DCT encoder and decoder structure were seen in Figure 5.1 and Figure 5.2. These same figures are appropriate for describing the basic structure utilized in the first three MPEG standards. The first frame in a video sequence (I-picture) is encoded in INTRA mode without reference to any past or future frames. At the encoder, the DCT is applied to each 8 x 8 luminance and chrominance block and, after output of the DCT, each of the 64 DCT coefficients is quantized (Q). The quantizer stepsize used to quantize the DCT-coefficients within a macroblock is transmitted to the receiver. Typically, the lowest DCT coefficient, which is the DC coefficient, is treated differently from the remaining AC coefficients. The DC coefficient corresponds to the average intensity of the component block and is encoded using a differential DC prediction method. The non-zero quantized values of the remaining DCT coefficients and their locations are then "zig-zag" scanned and run-length entropy coded using VLC tables.

Each subsequent frame of an MPEG encoder is coded using either inter-frame prediction (P-pictures) or bi-directional frame prediction (B-pictures). For both P and B pictures, only data from the nearest previously coded I or P frame is used for prediction. For coding a P-picture at frame N, the previously coded I or P picture, frame N-1, is stored in memory in both encoder and decoder. Motion compensation is performed on a macroblock basis. Only one motion vector is estimated between frame N and frame N-1 for a particular macroblock to be encoded. Each motion vector is coded and transmitted to the receiver.

B-pictures are coded using motion-compensated prediction based on the two nearest coded reference frames. These two reference frames are either an I picture and a P-picture, or two P-pictures. A B-picture is never used as a reference for the prediction of any other frame. The direction of prediction and relationships between the three MPEG picture types is shown in Figure 5.11. In order to suit the needs of diverse applications, the arrangement of the picture coding types within the video sequence is flexible. This arrangement or ordering of picture types is referred to as the Group of Pictures (GOP) structure. The GOP specifies the spacing and number of P and B pictures that are included between two I-pictures.

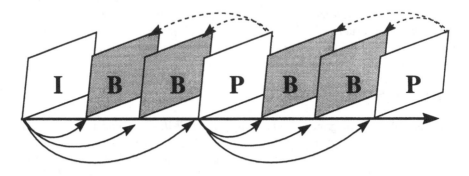

Figure 5.11: Group of Pictures in MPEG.

5.4.2 MPEG-1

As mentioned, the MPEG-1 video compression standard was targeted at multimedia CD-ROM applications. Utilizing the JPEG and H.261 standards as starting point, MPEG-1 developers focused on standardizing additional tools to enable new functionalities necessary to support this application. These tools include frame based random access of video, fast forward/fast reverse (FF/FR) searches through compressed bitstreams, reverse playback of video and a compressed bitstream that could be edited.

By focusing on applications that utilize a CD-ROM, developers of the MPEG-1 standard were able to fix many of the parameters associated with a communication channel, which are generally not known. For instance, based on this application, the channel bitrate for MPEG-1 was assumed to be 1.5 Mbps, which is the data transfer rate of a single spin CD-ROM. More importantly for error resilience, the channel error characteristics were also fixed to that of a CD-ROM.

CD-ROM error characteristics. The error characteristics associated with a CD-ROM are much less severe than those of other storage media, such as magnetic tapes. The main source of errors is the imperfections in the disk media. An imperfection as small as 2mm in diameter can cause a burst of up to 4000 errors to be injected into the bitstream [18].   Other sources of errors include fingerprints and scratches. To mitigate these errors, the data on a CD-ROM is encoded using techniques such as cross-interleave Reed-Solomon code (CIRC). This very powerful forward error correcting approach utilizes two Reed-Solomon encoders with an interleaver.   At the decoder, the burst errors are spread out by the deinterleaver, such that they appear as random errors to the Reed-Solomon decoders.   The Reed-Solomon decoders then correct these isolated bit errors and provide the video decoder with an essentially error-free bitstream.

MPEG-1 System Layer. The MPEG-1 system layer was developed in order to address the following problems: synchronization of decoded audio and video, buffer management to prevent overflow and underflow, random access start-up, and absolute time identification. To solve these problems, the MPEG-1 system layer utilizes time stamps to specify the decoding and display of audio and video at the decoder.   This simple approach allows a great deal of flexibility in the decoding of source material with varying frame rates and audio sample rates.   It also provides for the multiplexing of multiple simultaneous audio and video streams as well as data streams. However, other than a detecting an out of order time stamp, the MPEG-1 system layer is not designed to be error resilient.

MPEG-1 coding tools.   As mentioned, the video compression algorithm of MPEG-1 was built upon the existing standards of JPEG and H.261. Consequently, many of the basic tools found in MPEG-1, JPEG and H.261 are similar. This is apparent from the fact that the block diagrams of Figure 5.1 and Figure 5.2 accurately describe both H.261 and MPEG-1. In addition to the DCT, motion estimation/compensation, quantization and entropy encoding, MPEG-1 also organizes and processes the data as macroblocks.   Furthermore, the processed macroblocks are grouped into slices.   As discussed earlier, slices are very important for localizing an error or errors that have corrupted a video bitstream.

In addition to the video compression techniques of H.261, MPEG-1 also supports functionalities necessary for CD-ROM applications. Through modifications in the syntax structure of H.261, functionalities such as frame based random access of video, fast forward/fast reverse searches through compressed bit streams, reverse playback of video and compressed bit stream editing are all possible in MPEG-1. It is the addition of the GOP that enables these new functionalities. The user can arrange the picture types in a video sequence with a high degree of flexibility to suit diverse application requirements. As a general rule, a video sequence coded using I pictures only allows the highest degree of random access, FF/FR and also error resilience, but achieves only low compression. A sequence coded with a regular I picture update and no B pictures achieves moderate compression and a certain degree of random access, FF/FR, and error resilience functionality. Incorporation of all three pictures types, as depicted in Figure 5.11, may achieve high compression and reasonable random access, FF/FR, and error resilience functionality, but also increases the coding delay significantly. This delay may not be tolerable for applications such as video telephony or videoconferencing.

5.4.3   MPEG-2

MPEG-2 was developed to provide video coding solutions for applications not originally covered or envisaged by the MPEG-1 standard. Specifically, MPEG-2 was given the charter to provide video quality not lower than NTSC or PAL and up to ITU-T Rec. 601 quality. Emerging applications, such as digital cable TV distribution, networked database services via ATM, digital VTR applications and satellite and terrestrial digital broadcasting distribution were seen to benefit from the increased quality expected to result from the new MPEG-2 standardization phase. The target data rate for MPEG-2 is between four and nine Mbps, dependent upon the scene content and network capacity. For applications such as direct broadcast satellite, the data rate varies between three and six Mbps.

Also dependent on the network is the particular system layer to be used and whether error resilience tools will be required. Distribution of MPEG-2 over packet-based networks, where packet loss is a reality, is what forced the MPEG committee to begin to work on the problems associated with error prone channels. Emerging from this work are several modifications to both the compression and system layers of MPEG-1 which improve the performance of MPEG-2 over MPEG-1 when operating over an error prone channel.

MPEG-2 Systems Layer. To effectively distribute entertainment quality multimedia, the system layer for MPEG-2 must be able to simultaneously transmit a video bitstream along with more than two channels of audio. The MPEG-2 systems layer provides two different approaches to solving this problem. The first method called the Program Stream multiplexes the audio and video data in manner very similar to MPEG-1. The second approach called the Transport Stream (TS) uses a very different approach to multiplex the various data.

The Transport Stream is a packet-oriented multiplex designed to operate over ATM based networks. It has the ability to multiplex several streams of audio, video and data onto a single output channel. Presentation (i.e., display of the decoded video with audio) of the various bitstreams is synchronized through the use of time stamps, which are transmitted with each packet. The packetization of the different elementary streams is carried out in two steps. The first step is to packetize each elementary source data stream. Similar to the Adaptation Layer of H.223, these packets, which are called Packetized Elementary Stream (PES) packets, are of variable length. The header of a PES packet includes an optional CRC to allow for error detection within the header.

In the second step, the PES packets are segmented into smaller packets of 188 bytes in length. These smaller packets are called Transport Stream packets. The TS packets include a 4-byte header, which allows the decoder to send the received data to the correct elementary decoder (e.g., video or audio). The TS packets are transmitted over the channel by time multiplexing packets from different elementary streams. For ATM-based networks such as B-ISDN and many wireless W-LAN, the TS packets are further divided into ATM size packets of 48 bytes.

Error resilience of the MPEG-2 Transport Stream is greatly enhanced by the fact that the TS packets are of fixed length. At the system layer, loss of synchronization is not a problem. Furthermore, the ability of the system layer decoder to decode a TS packet is independent of whether or not any previous packets were corrupted.

MPEG-2 encoding tools. The video coding scheme used in MPEG-2 is again generic and similar to MPEG-1 and H.261. However, MPEG-2 also contains further refinements and special considerations for interlaced sources. One such interlaced compression tool is "Dual Prime", which utilizes new motion compensation modes to efficiently explore temporal redundancies between interlaced fields. New functionalities, such as error resilience, scalability and 4:2:2 formats were also introduced into MPEG-2.

MPEG-2 error resilience tools. As mentioned above, the MPEG-2 syntax supports error resilient modes. These modes were developed to help facilitate the use of MPEG-2 over ATM-based networks. However, these tools are of use in any error environment that includes isolated and burst errors. A slice structure, discussed previously, is also utilized in MPEG-2 for resynchronization of the decoder after an error or errors has corrupted the transmitted bitstream. The slice structure of MPEG-2 is more flexible than what is available in H.261, H.263 and MPEG-1. The length of a particular slice is dependent on the start of the next transport packet. This allows MPEG-2 to align the start of a slice with the start of an ATM packet. This is extremely efficient from both an overhead and error resilience point of view, since each ATM packet will contain only one slice start code. This helps localize a lost ATM packet to a single slice. This improved localization results in a smaller area to conceal.

To improve a decoder's chances of concealing a bit error, MPEG-2 allows for the transmission of additional motion vectors. These motion vectors are utilized to motion compensate the region of the previous frame which corresponds to the region affected by errors in the present frame. Since macroblocks which are predicted (this includes both P and B type prediction) already posses motion vectors, transmission of additional motion vectors is redundant. However, INTRA-macroblocks do not have motion vectors. Therefore, if an affected region is surrounded by Intra-macroblocks then it would be very difficult to perform motion-compensated concealment. To alleviate this problem, MPEG-2 allows motion vectors to be transmitted with INTRA-macroblocks, for use in error concealment.

Data partitioning is another error resilient tool provided by MPEG-2. This tool allows a straightforward separation of the encoded data for a macroblock into two layers. Layer 1 of this partition includes the macroblock address and control information, motion vectors and possibly the low frequency DCT coefficients. Layer 2 contains the remaining (high frequency) DCT coefficients. The layers are created within a slice and, depending on the type of macroblocks contained in that slice, the amount information within each layer can vary.

The information contained in Layer 1 is crucial to the decoding process and therefore of higher priority than the information contained in Layer 2. In order to provide reasonable error resilience, Layer 1 must be transmitted with fewer errors than Layer 2. In other words, a higher Quality of Service (QoS) channel must be assigned to Layer 1. Generally, this higher QoS is achieved by use of error correcting codes. By separating the bitstream into two distinct layers, data partitioning is very similar to a scalable bitstream. However, the major difference is that the quality of the decoded sequence resulting from Layer 1 is dependent on Layer 2. In other words, unlike a truly scalable bitstream, Layer 1 of a partitioned bitstream does not guarantee a minimum level of quality.

MPEG-2 Scalability. Scalability was originally intended to provide interoperability between different services and to support receivers with different display capabilities. Receivers either not capable of or willing to reconstruct the full resolution video can decode subsets of the layered bit stream to display video at lower quality (SNR), spatial or temporal resolution with lower quality. Another important purpose of scalable coding is to provide a layered video bit stream which is amenable for prioritized transmission. The main challenge here is to reliably deliver video signals in the presence of channel errors, such as cell loss in ATM based transmission networks or co-channel interference in terrestrial digital broadcasting.

For example, SNR scalability produces a bitstream with two layers: a base layer and an enhancement layer. The base layer, when decoded, results in a sequence with the same spatial resolution as the original source, but with lower perceived quality to the viewer. When combined with an enhancement layer, the quality of the decoded sequence is greatly improved. The dependencies and importance of each layer are the same as those discussed for H.263 Scalability mode.

This natural prioritizing of the bitstream is very useful for providing error resilience in MPEG-2. For instance, the higher priority base layer is typically transmitted over a higher QoS channel than the enhancement layer. This translates to fewer errors corrupting the base layer as opposed to the enhancement layer. If a region of the enhancement layer is corrupted by errors and lost, it can be concealed utilizing the base layer in the corresponding region.

In another example, a high degree of error resilience can be achieved with temporal scalability by encoding the base-layer using the same spatial resolution but only half the temporal resolution of the source [19]. The remaining frames are encoded as the enhancement-layer. Again as discussed above, to take full advantage of the error the error resilience capabilities of scalability, the base-layer is given a higher priority or QoS identifier. The enhancement layer, considered to be lower priority, has a greater chance of being corrupted in transmission and therefore lost. If a frame of the enhancement-layer is lost, it is concealed by using frame repetition of the base layer.

5.4.4 MPEG-4

MPEG-4 is the first video compression standard which addresses the efficient representation of visual objects of arbitrary shape. It is the goal of MPEG-4 to provide the functionalities necessary to support content-based functionalities. Furthermore, it will also support most functionalities already provided by MPEG-1 and MPEG-2, including the provision to efficiently compress standard rectangular sized image sequences at varying levels of input formats, frame rates, bitrates, and various levels of spatial, temporal and quality scalability. A second major goal of MPEG-4 is to provide the functionality of "Universal Accessibility". In other words, MPEG-4 addresses the problem of accessing audio and video information over a wide range of storage and transmission media. In particular, due to the rapid growth of mobile communications, it is extremely important that access to audio and video information be available via wireless networks.

MPEG-4 encoding tools. The coding of conventional video by MPEG-4 is achieved utilizing tools which are similar to those available in both MPEG-1 and MPEG-2. Specifically, MPEG-4 utilizes motion prediction/compensation followed by texture coding as shown in Figure 5.1. For the content-based functionalities, where the input video sequence may be of arbitrary shape and location, MPEG-4 utilizes additional tools such as shape coding. The shape of an object is represented in MPEG-4 in one of two ways. The first is as a binary mask. This method is utilized to encode the contours, or boundary shape of an object. When transparency information needs to be sent, an alpha mask is encoded and transmitted. In both cases, motion-compensated prediction and texture coding utilizing a DCT is used to encode the interior of an object.

One of the other interesting differences from MPEG-1 and MPEG-2 is the use of global motion compensation based on the transmission of "sprites". A sprite is a possibly large still image, often describing a panoramic background. For each consecutive image in a sequence, only 8 global motion parameters describing camera motion are coded to reconstruct the object.

MPEG-4 universal accessibility. MPEG-4 provides error robustness and resilience to allow accessing image or video information over a wide range of storage and transmission media. In particular, due to the rapid growth of mobile communications, it is extremely important that access is available to audio and video information via wireless networks. This implies a need for useful operation of audio and video compression algorithms in error-prone environments at low bit-rates (i.e., less than 144 Kbps).

The error resilience tools developed for MPEG-4 can be divided into three major areas. These areas or categories include resynchronization, data recovery, and error concealment. It should be noted that these categories are not unique to MPEG-4, but instead have been used by many researchers working in the area error resilience for video. It is, however, the tools contained in these categories that are where MPEG-4 makes its contribution to the problem of error resilience. In this section, the MPEG-4 resynchronization and data recovery tools will be discussed.

MPEG-4 resynchronization. Resynchronization tools, as the name implies, attempt to enable resynchronization between the decoder and the bitstream after a residual error or errors have been detected. Generally, the data between the synchronization point prior to the error and the first point where synchronization is reestablished, are discarded. If the resynchronization approach is effective at localizing the amount of data discarded by the decoder, then the ability of other types of tools which recover data and/or conceal the effects of errors is greatly enhanced.

The resynchronization approach adopted by MPEG-4, referred to as a packet approach, is similar to the resynchronization structure utilized by the Slice Structured mode of H.263. The video packet approach adopted by MPEG-4, like the adaptive slice of MPEG-2 and Slice Structured mode H.263, is based on providing periodic resynchronization markers throughout the bitstream. In other words, the length of the video packets are not based on the number of macroblocks, but instead on the number of bits contained in that packet. If the number of bits contained in the current video packet exceeds a predetermined threshold, then a new video packet is created at the start of the next macroblock.

A resynchronization marker is used to distinguish the start of a new video packet. This marker is distinguishable from all possible VLC codewords as well as the picture, or as it is called in MPEG-4 a Video Object Plane (VOP), start code. Header information is also provided at the start of a video packet. Contained in this header is the information necessary to restart the decoding process and includes the macroblock number of the first macroblock contained in this packet and the quantization parameter necessary to decode that first macroblock. The macroblock number provides the necessary spatial resynchronization while the quantization parameter allows the differential decoding process to be restarted. It should be noted that when utilizing the error resilience tools within MPEG-4, some of the compression efficiency tools are modified. For example, all predictively encoded information, including advanced predictive I macroblock codes of MPEG-4, must be confined within a video packet to prevent the propagation of errors.

In conjunction with the video packet approach to resynchronization, a second method called fixed interval synchronization has also been adopted by MPEG-4. This method requires that VOP start codes and resynchronization markers appear only at legal fixed interval locations in the bitstream. This helps to avoid the problems associated with start code emulations. This is because, when fixed interval synchronization is utilized, the decoder is only required to search for a VOP start code at the beginning of each fixed interval. The fixed interval synchronization method extends this approach to any predetermined interval.

MPEG-4 data partitioning. In recognizing the need to provide enhanced concealment capabilities, MPEG-4 also utilizes data partitioning. This data partitioning is achieved by separating the motion and macroblock header information away from the texture information. It should be noted that, when present, shape data is also partitioned with the motion data away from the texture data.

This approach requires that a second resynchronization marker, the "motion marker" be inserted between motion and texture information. Figure 5.12 illustrates the syntactic structure of the data partitioning mode. If the texture information is lost, this approach permits the use of motion information to conceal these errors.

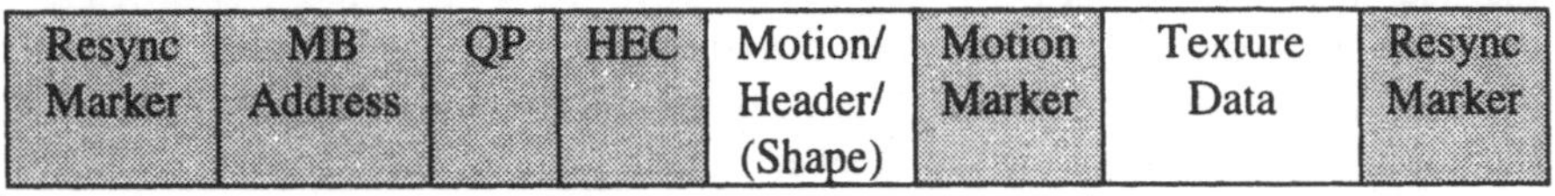

Resync Marker	MB Address	QP	HEC	Motion/ Header/ (Shape)	Motion Marker	Texture Data	Resync Marker

Figure 5.12: Data Partitioning in MPEG-4.

MPEG-4 data recovery. After synchronization has been reestablished, data recovery tools attempt to recover data that, in general, would be lost. These tools are not simply error correcting codes, but instead techniques that encode the data in an error resilient manner. For instance, one particular tool is in MPEG-4 Video is Reversible Variable Length Codes (RVLC). In this approach, the variable length codewords are designed such that they can be read both in the forward as well as the reverse direction.

An example illustrating the use of an RVLC is given in Figure 5.13. Generally, in a situation such as this, where a burst of errors has corrupted a portion of the data, all data between the two synchronization points would be lost. However, as shown in Figure 5.13, an RVLC enables some of that data to be recovered. It should be noted that the parameters, QP and HEC shown in Figure 5.13, represent the fields reserved in the video packet header for the quantization parameter and the header extension code, respectively. If the HEC is enabled, then additional control information is included in the header. This additional control information can include items such as the temporal reference.

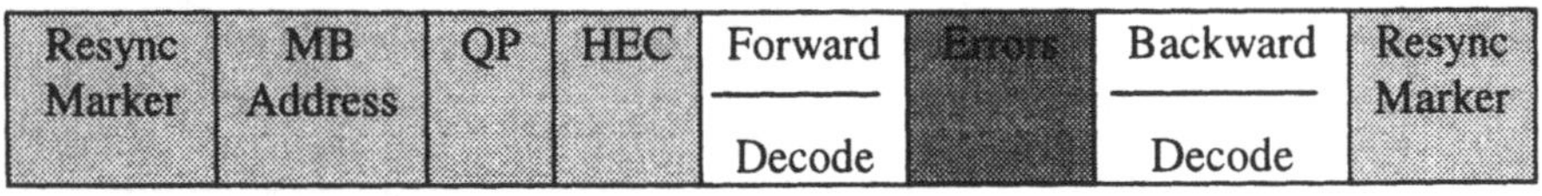

Resync Marker	MB Address	QP	HEC	Forward ——— Decode	Errors	Backward ——— Decode	Resync Marker

Figure 5.13: Example of Reversible Variable Length Code.

5.5   ROBUST VIDEO DECODERS

For bandwidth-limited real-time applications, and applications which cannot tolerate significant delay, it is often necessary to accept some level of errors in the video bitstream. In other words, it may not be possible, or desirable to spend a large percentage of available bits on forward error correction, or to suffer the delay of ARQ or bit interleaving. However, it may still be possible to provide an entirely acceptable quality of service by simply using intelligent decoder algorithms.

There are no specific requirements within the standards to specify how a video decoder should detect and conceal errors, but there are a number of recommended procedures that are facilitated by the particular syntax of each standard. The three most prevalent approaches to providing for robustness within a video decoder are: containment, recovery and concealment. In this section, some typical methods provided by H.263 and MPEG-4 for each of these approaches are discussed and demonstrated.

5.5.1   Containment

The first goal of any error robust video decoder should be to localize or "contain" any errors that do get through the demultiplexer into the video decoder. As was discussed previously, the generous placement of resynchronization bits within the video bitstream enables a decoder to localize the effect of errors spatially to within these resynchronization points.

Utilizing resynchronization information. The use of resynchronization points within the bitstream is very important to enabling concealment. When an error is detected, a video decoder must stop decoding and parse the bitstream for the next resynchronization point. In most of the cases discussed in this chapter, this is easily accomplished by correlating the incoming bitstream with the known synchronization flag bits. A decoder can either look for an exact match of that flag or a match within some level of confidence before resetting and continuing the decoding process.

The important functionality of the decoder that is needed to effectively utilize these resynchronization points is good error detection. It was mentioned previously that detecting errors is not always easy when reading codewords from a nearly complete VLC tree. However, the parameters in the resynchronization headers themselves provide a large portion of the information needed to detect errors. When a bad codeword goes unnoticed by the decoder, the macroblock count, or the spatial address of the pixels that the decoder thinks correspond to the current position in the bitstream, can be corrupted. The GOB or slice headers always indicate the absolute count of the next macroblock in the bitstream. If this does not match the expected count in the decoder, an error is detected and can be treated accordingly.

The frequency of resynchronization information in the bitstream has also been discussed earlier. This is a choice of the encoder. The more resilient standards support a wide range of levels of resynchronization information. If the encoder has any basic knowledge about typical channel conditions, the frequency of resynchronization points can be controlled to match these conditions. For example, if the average error burst length of the channel is known to be 2000 bits, it is not very efficient to send resynchronization points every 100 bits. Although bursts could be better localized, most of the time the resynchronization overhead will be wasted, resulting in worse compression efficiency for the error free periods of video.

5.5.2 Recovery

Once an error has been localized to within the boundaries of some resynchronization points, it may be possible to recover some the information in that segment. The amount of possible recovery is dependent upon the video syntax, and standards like MPEG-4 are best at providing for data recovery. In MPEG-4, this is accomplished through data partitioning and RVLCs.

In order to take advantage of RVLCs, an MPEG-4 decoder must have a convention for discarding and keeping certain information in a video packet after an error has been detected. For example, one of the strategies recommended by the MPEG-4 specification is seen in Figure 5.14, where the following definitions are used:

- L = Total number of bits for DCT coefficients in the video packet,

- N = Total number of macroblocks (MBs) in the video packet,

- $L1$ = Number of bits which can be decoded in the forward direction,

- $L2$ = Number of bits which can be decoded in the backward direction,

- $N1$ = Number of MBs which can be decoded completely in the forward direction,

- $N2$ = Number of MBs which can be decoded completely the backward direction,

- $f\_mb(S)$ = Number of decoded MBs when S bits can be decoded in the forward direction,

- $b\_mb(S)$ = Number of decoded MBs when S bits can be decoded the backward direction,

- T = Threshold (90 is the MPEG-4 recommended value).

This strategy is applied when $L1+L2 < L$ and $N1+N2 < N$. Here, MBs of $f\_mb(L1\text{-}T)$ from the beginning and MBs of $b\_mb(L2\text{-}T)$ from the end of the video packet are used. In Figure 5.14, the MBs of the dark region are discarded. MPEG recommends additional strategies for other cases of $L1$ and $L2$, and for the treatment of INTRA coded macroblocks in the bitstream17. INTRA MBs are best left concealed when there is uncertainty about their validity. This is because the result of displaying an INTRA MB that does contain an error can substantially degrade the quality of the video.

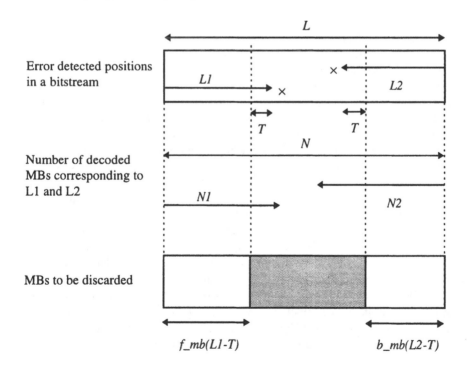

Figure 5.14: One decoding strategy for MPEG-4 bitstream when error are detected and RVLCs are present.

5.5.3 Concealment

After all of the data within an erroneous segment which can be recovered has been obtained, the final region requiring concealment is identified. Such a region would correspond to the dark region of Figure 5.14. This region typically consists of a number of contiguous macroblocks in the bitstream which either were not decoded, or cannot be displayed with any confidence based on the detected errors. These macroblocks can be concealed in any number of ways, however, the methods used are always outside of the scope of the standards.

Figure 5.15 shows the effects of containment and concealment applied to 3 different H.263 bitstreams using different amounts of resynchronization information. Each of the three bitstreams of "Carphone" were encoded at CIF resolution at 46.4 Kbps, and subjected to the identical error pattern using an average burst length of 10ms [20]. The detected errors were localized to the smallest possible region in each case based on the resynchronization segment size. The bad macroblocks were concealed by replacing them with the corresponding macroblocks from the previous decoded frame. Notice the effects of increasing amounts of resynchronization points on the reconstructed image quality. These differences are solely due to the ability to better conceal small localized regions within the picture.

MPEG-4 provides the tool of data partitioning, as mentioned previously. When coupled with a smart decoder, the idea of separate motion vectors and texture information within one resynchronization segment of the bitstream is very important. If part of the motion vectors can be recovered with confidence, it is then possible to conceal data by not using the macroblock at the same spatial position in the previous frame, but rather by using "motion-compensated" concealment.

Motion-compensated concealment can lead to better decoded video quality in the presence of any amount of scene activity. This is best demonstrated by the results presented in Figure 5.16 for a sequence taken from the MPEG-4 test suite of video sequences, "Silent Voice". The graph shows the PSNR at each decoded frame, averaged over 100 runs, using an error prone channel simulation of 10 ms bursts with an average BER of 10^{-3}. These results, taken from a core experiment within MPEG-4, show that the use of motion-compensated concealment vs. non-motion-compensated macroblock concealment is quite effective, reflected in the 1 dB of objective gain seen in this example [21]. The PSNR naturally decreases over time in both cases because of the introduction and propagation of errors and concealed information. The use of data partitioning, however, allows more bits to be kept and used by the decoder in the motion-compensated case, which results in a better reconstruction of the scene, and a better PSNR.

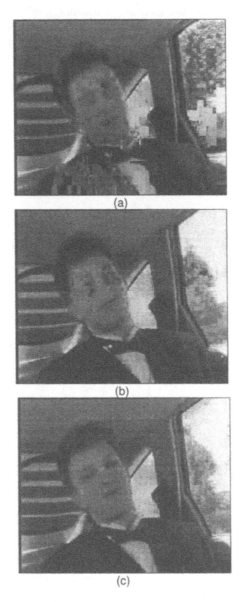

(a)

(b)

(c)

Figure 5.15: Decoded frame from H.263 compressed sequence "Carphone", coded at 46.4 Kbps, CIF resolution. Bitstreams were corrupted by identical random burst errors of average length 10ms. Resynchronization points coded at: (a) frame headers only; (b) every other GOB header; (c) every 512 bits.

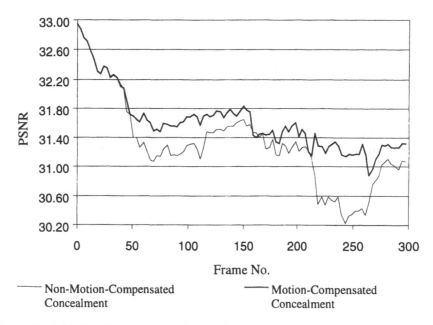

Figure 5.16: Motion-Compensated Concealment vs. Non-Motion-Compensated Macroblock Concealment; "Silent Voice" sequence, 100 runs of 10^{-3} avg. BER, 10 ms bursts.

This concept is also illustrated by the use of motion vectors with INTRA coded macroblocks for concealment in the face of packet loss in MPEG-2 coding [22]. It is also possible to provide error concealment directly in the transform domain in the case of minor random errors. For example, an objective function describing inter-sample variations at the boundaries of lost blocks has been used to recover corrupted DCT coefficients by solving a linear equation [23]. This illustrates the point that, in general, much of the research in the area of image and video restoration is facilitated by the syntax of the standards discussed here.

5.6 CONCLUSIONS

Error resilience is emerging as an important functionality considered by all of the major video standards in the last several years. Its importance stems from an increasing ability to communicate digital information in wireless and packet-based environments. This comes as the next generations of digital cellular technology make their way to the marketplace, potentially providing enough bandwidth to the individual user that wireless multimedia communication can become a reality. Additionally, the rapid expanse of network communications and the use of non-guaranteed delivery protocols also have driven the quest for error resilient standards.

Most of the time, heavily compressed video data cannot handle the impact of bit errors that come with degraded channels. However, given the appropriate sets of tools for combating the effects of errors, it is possible to achieve good quality video communication. The syntax of the standards discussed in this chapter have specific elements dedicated to preserving the quality of decoded video in the presence of errors. These elements provide the means for a decoder to accomplish three important tasks: containment, recovery, and concealment. With good error detection in a standardized video decoder, it is possible to localize and contain the errors using the many methods for resynchronization described here. Within the localized area, lost data can often be recovered as a result of data partitioning. Finally, the erroneous segments of the picture can be concealed using either motion-compensated or non-motion-compensated block concealment. The ability to use these tools and operate with some errors at the receiver frees encoders to spend more bits on compression efficiency, and not on excessive error correction and retransmission overhead which can be extremely inefficient in terms of bits and overall system delay.

References

[1] B. Sklar, "Rayleigh fading channels in mobile digital communications systems part I: characterization," *IEEE Communications Magazine,* vol. 35, pp. 90-100, July, 1997.

[2] ITU-T Recommendation H.320, "Narrow-band visual telephone systems and terminal equipment," March 1996.

[3] ITU-T Recommendation H.221, "Frame structure for a 64 to 1920 kbit/s channel in audiovisual teleservices," July, 1995.

[4] ITU-T Recommendation H.261," Video codec for audiovisual services at p x 64 kbit/s," March 1993.

[5] ITU-T Recommendation H.324, "Terminal for low bit rate Multimedia Communication," March 1996.

[6] ITU-T Recommendation H.324/Annex C, "Multimedia telephone terminals over error prone channels," Jan. 1998.

[7] ITU-T Recommendation H.223, "Multiplexing protocol for low bitrate multimedia communication," March 1996.

[8] ITU-T Recommendation H.223/Annex A, "Multiplexing protocol for low bitrate mobile multimedia communication over low error-prone channels," Jan. 1998.

[9] M. R. Banham and T. Kadir, "Verification Tests for Synchronization Flag in H.223 Annexes A and B", ITU-T SG16/Q11, Doc. Q11-C-20, Eibsee, Germany, Dec. 2-5, 1997.

[10] T. Kadir, M. R. Banham, and J. C. Brailean, "Core Experiment Results for Multi-Level Error Resilient Extensions for H.223", ITU-T SG16/Q11, Doc. Q11-B-24, Sunriver, OR, Sept. 8-12, 1997.

[11] ITU-T Recommendation H.223/Annex B, "Multiplexing protocol for low bitrate multimedia communication over moderate error-prone channels," Jan. 1998.

[12] ITU-T Recommendation H.223/Annex C, "Multiplexing protocol for low bitrate multimedia communication over highly error-prone channels," Jan. 1998.

[13] J. Hagenauer, "Rate-Compatible Punctured Convolutional Codes (RCPC Codes) and their Applications," *IEEE Trans. Communications*, vol. 36, pp. 389-400, April 1988.

[14] ITU-T Recommendation H.263, Version 2, "Video coding for low bitrate communication," Jan. 1998.

[15] ISO/IEC 11172-2, "Information technology-coding of moving pictures and associated audio for digital storage media at up to about 1.5 Mbits/s: Part 2 Video," Aug. 1993.

[16] ITU-T Recommendation H.262 | ISO/IEC 13818-2, "Information technology - generic coding of moving pictures and associated audio information: video," 1995.

[17] ISO/IEC 14496-2, "Information technology-coding of audio-visual objects: Visual," Committee Draft, Oct. 1997.

[18] B. Sklar, *Digital Communications: Fundamentals and Applications.* Englewood Cliffs, NJ: Prentice Hall, 1988.

[19] ITU-T Recommendation H.262 I ISO/IEC 13818-2, "Information technology - generic coding of moving pictures and associated audio information: video," 1995.

[20] T. Miki, T. Kawahara, and T. Ohya, "Revised Error Pattern Generation Programs for Core Experiments on Error Resilience," ISO/IEC JTC1/SC29/WG11 MPEG96/1492, Maceio, Brazil, Nov. 1996.

[21] M. R. Banham and R. Dean, "Results for Core Experiment E5: Error Concealment and Resynchronization Syntax for Separate Motion Texture Mode," ISO/IEC JTC1/SC29/WG11 MPEG96/1476, Maceio, Brazil, Nov. 1996.

[22] R. Aravind, R. Civanlar, and A. R. Reibman, "Packet loss resilience of MPEG-2 scalable video coding algorithms, " IEEE Trans. Circuits Syst. Video Technol., vol. 6, pp. 426-435, Oct. 1996.

[23] J. W. Park, J. W. Kim, S. U. Lee, "DCT coefficient recovery-based error concealment technique and its application to the MPEG-2 bit stream error," IEEE Trans. Circuits Syst. Video Technol., vol. 7, pp. 845-854, Dec. 1997.

6 ERROR-RESILIENT STANDARD-COMPLIANT VIDEO CODING

Bernd Girod and Niko Färber

Telecommunications Laboratory
University of Erlangen-Nuremberg
Cauerstrasse 7, 91058 Erlangen, Germany
Phone: +49-9131-857101, Fax: +49-9131-858849
{girod, faerber}@nt.e-technik.uni-erlangen.de

Abstract:

In this chapter we review and compare two approaches to robust video transmission that can be implemented within the H.263 video compression standard. The focus of this chapter is on channel adaptive approaches that rely on a feedback channel between transmitter and receiver carrying acknowledgment information. Based on the feedback information, rapid error recovery is achieved by intra refresh of erroneous image regions. The consideration of spatial error propagation provides an additional advantage. Though the average gain is less than 0.3 dB, annoying artifacts can be avoided in particularly unfavorable cases. The feedback messages that need to be defined outside the scope of H.263 are supported in the ITU-T Recommendation H.324 that describes terminals for low bit-rate multimedia communication. In order to investigate the influence of error concealment we provide simulation results for concealment with the zero motion vector and the true motion vector as a lower and an upper bound respectively. Experimental results with bursty bit error sequences simulating a wireless DECT channel at various Signal to Noise Ratios are presented in order to compare the different approaches. We are using a simple Forward Error Correction (FEC) scheme on the forward channel while assuming error-free transmission and a fixed delay of 100 ms for the backward channel. For the comparison of picture quality we distinguish between distortion caused by coding and distortion caused by transmission errors. For the second kind of distortion we derive a model that is verified by experimental results from an H.263 decoder. Whenever appropriate, a discussion of system and complexity issues is included.

6.1 INTRODUCTION

ITU-T Recommendation H.324 describes terminals for low bit-rate multimedia communication, that may support real-time voice, data, and video, or any combination, including videotelephony. Because the transmission is based on V.34 modems operating over the widely available public switched telephone network (PSTN), H.324 terminals are likely to play a major role in future multimedia applications. In fact, an increasing number of H.324 terminals is already being implemented and purchased by various companies and vendors. One important reason for this success was the availability of the H.263 video coding standard [1], that achieves acceptable image quality at less than 32 kbps. Other Recommendations in the H.324 series include the H.223 multiplex, H.245 control protocol, and G.723 audio codec.

As mobile communication becomes a more important part of daily life, the next step is to support mobile multimedia communication. Recognizing this development, the ITU-T started a new "Mobile" Ad Hoc Group (AHG) in 1994 to investigate the use of H.324 in mobile environments. In the following paragraphs we will give a short summary of the main issues discussed in the Mobile AHG until 1997 that are related to robust video transmission. Since we focus on standard compatible extensions in this chapter, the reader is referred to [2] for a more general review of error control and concealment for video communication.

During the work of the Mobile AHG, it turned out that one major requirement for a mobile H.324 terminal is its ability to interwork with terminals connected to the PSTN at a reasonable complexity and low delay. This resulted in the decision to use the H.263 video codec and G.723 audio codec unchanged, because transcoding was considered to be too complex. With this decision, many promising proposals became obsolete. For example, the reordering of the H.263 bit stream into classes of different sensitivity (data partitioning) has been proposed to enable the use of unequal error protection [3] [4]. Though this approach is known to be effective [5] [6] [7], the parsing and re-assembling of the bit stream with added error protection cannot be implemented in a low complexity, low delay interworking unit.

Two other useful proposals, which do not have the drawback of high complexity, could not be adopted by the ITU-T, because they would have required minor changes in the H.263 bit stream syntax. Because H.263 was already "frozen" in January 1996, neither the SUB-VIDEO approach [8] nor the NEWPRED approach [9] have been included in H.263 for increased error robustness. However, both approaches are now included in a slightly modified and extended way in version 2 of the H.263 video coding standard, which is informally known as H.263+ and was adopted by ITU-T in February 1998. The NEWPRED approach is covered in Annex N (Reference Picture

Selection), while the SUB-VIDEO approach is covered in Annex R (Independent Segment Decoding). Future versions may also include data partitioning as already available in MPEG-4.

Given the above restriction on changes in H.263, standard compatible extensions for robust video transmission, as compared in this chapter, were a valuable enhancement of H.324 for the use in mobile environments. During the work in the Mobile AHG the authors proposed the Error-Tracking approach that is presented in the following sections. Because of its compatibility, no technical changes were needed in H.263. However, an informative appendix (Appendix II) was added to explain the basic concept of the Error-Tracking approach. In addition, minor extensions of the H.245 control standard were necessary to include the additional control message "videoNotDecodedMBs". Because the approach was developed in close relationship to the Mobile AHG, we will base our description on H.263. However, it should be noted that the approach can also be used with other video coding standards like H.261 or MPEG-2. Previous publications of our work on this approach include [10], [11], and [12].

This chapter is organized as follows. In Section 6.2 we discuss why residual errors are difficult to avoid in mobile environments for real-time transmission and why encoded video is particular sensitive to residual errors. After a brief review of hybrid video coding we discuss the effect of transmission errors in more detail in Section 6.3. In this section we also discuss error concealment and provide first simulation results that demonstrate the problem of error propagation. In Section 6.4 we derive a model that explains the propagation of error energy over time and verify it with experimental results. In particular, the influence of spatial filtering and INTRA updates are investigated. The Error-Tracking and Same-GOB approaches that use feedback information to effectively mitigate the effects of errors are described in Section 6.5. In Section 6.6, a low complexity algorithm for real-time reconstruction of spatio-temporal error propagation is described that can be used for the Error-Tracking approach. Experimental results demonstrate the performance of the investigated approaches for a simulated transmission over a wireless DECT channel in Section 6.7.

6.2 MOBILE VIDEO TRANSMISSION

Many existing mobile networks cannot provide a guaranteed quality of service, because temporally high bit error rates cannot be avoided during fading periods. Transmission errors of a mobile communication channel may range from single bit errors up to burst errors or even a temporal loss of signal. Those varying error conditions limit the effective use of Forward Error Correction (FEC), since a worst case design leads to a prohibitive amount of overhead. This is particularly true, if we have to cope with limited bandwidth requirements. Also note that the use of interleaving is constrained by low delay requirements. Closed-loop error control techniques like Automatic Repeat on reQuest (ARQ) have been shown to be more effective than FEC [13] [14] [15].

However, retransmission of corrupted data frames introduces additional delay, which is critical for real-time conversational services. Given an upper bound on the acceptable maximum delay, the number of retransmissions is mainly determined by the round-trip delay of data frames. For networks like DECT (Digital European Cordless Telephony), where data frames are sent and received every 10 ms, several retransmissions may be feasible. On the other hand, retransmission of video packets over a satellite link would introduce a prohibitively long delay. Therefore, residual errors are typical for real-time transmission in mobile environments even when FEC and ARQ are used in an optimum combination.

In the presence of residual errors, additional robustness is required because the compressed video signal is extremely vulnerable against transmission errors. Low bit-rate video coding schemes rely on INTER coding for high coding efficiency, i.e., they use the previous encoded and reconstructed video frame to predict the current video frame. Due to the nature of predictive coding, the loss of information in one frame has considerable impact on the quality of the following frames.

The severeness of residual errors can be reduced, if error concealment techniques are employed to hide visible distortion as well as possible [16], [17], [18], [19], [20]. However, even a sophisticated concealment strategy cannot totally avoid image degradation and the accumulation of several small errors can also result in poor image quality. To some extent, the impairment due to a transmission error decays over time due to leakage in the prediction loop. However, the leakage in standardized video decoders like H.263 is not very strong, and quick recovery can only be achieved when image regions are encoded in INTRA mode, i.e., without reference to a previous frame.

Summarizing the above paragraphs, we note that residual errors cannot be totally avoided for real-time transmission in mobile environments, and that error propagation is an important problem in mobile video transmission. To illustrate the second statement, we investigate the effects of error propagation in the framework of the H.263 video coding standard in the following section.

6.3 ERROR PROPAGATION IN HYBRID VIDEO CODING

We base our investigation of error propagation on the H.263 video compression standard as a typical representative of a hybrid video codec. Throughout this chapter we do not consider any optional coding modes of H.263. In this section we review the basic concept of H.263 and introduce some basic terms and signals that are used in the following to describe the performance of hybrid video coding in the presence of transmission errors. After we define the error concealment strategies in section 6.3.1 we present first simulation results in section 6.3.2 to demonstrate the typical impairment caused by error propagation.

The basic concept of H.263 is a temporal interframe prediction exploiting temporal redundancy followed by intraframe encoding of the residual prediction error. The temporal prediction is based on block-based motion estimation and compensation, while a discrete cosine transform (DCT) is used in the intraframe encoder to exploit spatial redundancy and adaptively reduce spatial resolution. Because predictive coding and transform coding are combined in one scheme it is often referred as hybrid.

The bit stream syntax of H.263 is based on the hierarchical video multiplex, that sub-divides each picture into logical layers with a specific functionality. In the following we focus on QCIF resolution (176×144 pel) which is the most common source format for bit-rates below 64 kbps. For QCIF, pictures are sub-divided into 9 Group Of Blocks (GOB, 176×16 pel), each consisting of 11 macroblocks (MB, 16×16 pel). One MB consists of 6 Blocks (8×8 pel), 4 for luminance and 2 for chrominance, respectively.

Because the multiplexed bit stream consists of variable length code words, a single bit error may cause a loss of synchronization and a series of invalid code words at the decoder. Even if resynchronization is achieved quickly, the following code words are useless if the information is encoded predictively like, e.g., motion vectors in H.263. Therefore H.263 supports optional start codes for each GOB that guarantee fast resynchronization. Furthermore, any dependencies from previous information in the current frame are avoided when the start code is present. The purpose of the GOB-Layer is therefore closely related to error robustness.

On the MB-Layer two basic modes of operation can be selected for each MB. In the INTER mode, motion compensated prediction is utilized on the basis of 16×16 blocks. The residual prediction error is encoded in the Block-Layer using a 8×8 DCT. If the INTRA mode is selected for a specific MB, no temporal dependency from previous frames is introduced and the picture is directly DCT coded. The choice between INTRA and INTER mode is not subject of the recommendation and may be selected as part of the coding control strategy. During normal encoding, however, the INTER mode is the preferred mode because of its significant advantage in coding efficiency.

To understand the overall performance of hybrid video coding in the presence of transmission errors, consider the simplified codec structure in Fig. 6.1 where the intraframe encoder is omitted. At the encoder, the original signal $o[x, y, t]$ is predicted by a motion compensated previous frame and the prediction error $e[x, y, t]$ is transmitted over the channel. Let the channel be error free except for $t = 0$, where the signal $u[x, y]$ is added to the prediction error $e[x, y, 0]$. Due to the recursive DPCM structure of the decoder the received signal $r[x, y, t]$ is not only erroneous for $t = 0$ but also for the following frames. Due to motion compensated prediction the error may also propagate spatially from its original occurrence. Typically, the error will also decrease its energy over time due to leakage in the DPCM loop as will be discussed in detail in section 6.4. All those effects are schematically illustrated in Fig. 6.2 for the loss of one GOB.

Note that there are two different types of errors to be considered. Referring to Fig. 6.1, we can define the distortion measure $D_c = E\{(o-c)^2\}$, i.e., the distortion caused by coding and $D_v = E\{(c-r)^2\}$, i.e., the distortion caused by transmission errors.

180

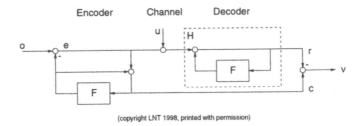

Figure 6.1   Simplified codec structure

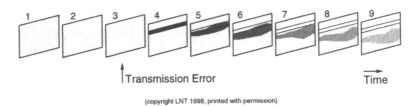

Figure 6.2   Spatio-temporal error propagation

Typically, D_c is the result of small quantization errors that are uniformly distributed over the whole frame, while D_v is dominated by strong errors that are concentrated in a small spatio-temporal part of the image. Assume that both error signals are zero-mean and statistically independent, in which case the above distortions are identical to the variance σ_{o-c}^2 and σ_v^2 of the signals $(o-c)$ and v respectively. Furthermore, the overall distortion after transmission $D_r = E\{(o-r)^2\}$ can be expressed as $D_r = D_c + D_v$ under this assumption. This illustrates that the overall distortion D_r combines two very different error types that are likely to be perceived differently. Therefore, especially when averaged over a whole sequence, the evaluation of D_r has to be done carefully. Because an average distortion measure is more useful for the same kind of errors, we try to match D_c for two approaches that shall be compared and use D_r and/or D_v as a performance measure.

6.3.1   Error Concealment

The error signal $u[x, y]$ that is introduced in the decoder as a result of transmission errors is the residual error after error concealment. For practical implementations it is very important to minimize this error. This can be achieved by reducing the erroneous image region by early resynchronization of the bit-stream parser, e.g., by inserting start codes for each GOB. In MPEG-4 more advanced techniques are provided based on reversible VLC codes in combination with data partitioning [21] [22]. On the other hand the introduced error $u[x, y]$ can be minimized by improving the error concealment for erroneous image regions.

In this chapter we will compare two approaches to error concealment that will serve as a lower and an upped bound. Both approaches operate on a GOB basis and therefore rely on the insertion of start codes for each GOB. GOBs are only decoded if received correct completely. Note that erroneous GOBs can be detected by using a cyclic redundancy check (CRC) or by the detection of syntax violations (e.g. invalid VLC codes, more than 64 DCT coefficients, wrong number of MBs per row, etc.).

In the first approach, that has also been proposed in the Mobile AHG [23] the image content of corrupted GOBs is replaced with data from the previously decoded frame. This corresponds to motion compensated concealment with the zero motion vector ("zero-MV concealment"). This simple approach has been shown to yield good results for sequences with little motion [16]. However, severe distortions are introduced for image regions containing heavy motion. In the following it will serve as a lower bound for the performance of error concealment.

More sophisticated and complex error concealment techniques such as described in [16], [17], [18], [19], [20], could be used as well. Instead, we will use motion compensated concealment with the true motion vector as the second approach ('true-MV concealment'). Though this approach cannot be implemented in a real system, it provides an upper bound for the performance of (motion compensated) concealment.

6.3.2 Signal-to-Noise Ratio Loss

For all simulation results in this chapter we express the distortion caused by transmission errors as a loss of picture quality compared to the error-free case. For easy interpretation we use the difference of PSNR values for corresponding frames, defined as

$$
\begin{aligned}
\Delta PSNR[t] &= 10\log\frac{255^2}{D_r[t]} - 10\log\frac{255^2}{D_c[t]} \\
&= 10\log\frac{D_c[t]}{D_r[t]},
\end{aligned}
\tag{6.1}
$$

where $D_r[t]$ and $D_c[t]$ are the mean squared errors of the received signals r and the coded signal c with respect to the original signal o as defined above. We either provide results for individual frames t or averaged results for a whole sequence. When the variance $\sigma_v^2[t]$ of the error signal v is given for each time step, we can formulate (6.1) as

$$
\begin{aligned}
\Delta PSNR[t] &= 10\log\frac{255^2}{\sigma_{o-c}^2+\sigma_v^2[t]} - 10\log\frac{255^2}{\sigma_{o-c}^2} \\
&= 10\log\frac{\sigma_{o-c}^2}{\sigma_{o-c}^2+\sigma_r^2[t]},
\end{aligned}
\tag{6.2}
$$

where σ_{o-c}^2 is set to correspond to the average picture quality after coding. This assumes that the error signal caused by transmission errors (v) and the error signal caused by coding ($o - c$) are zero-mean and statistically independent.

Fig. 6.3 shows the loss of picture quality ($\Delta PSNR$) after the loss of one GOB when the zero-MV concealment strategy is used. The QCIF sequence *Foreman* is coded at 100 kbps and 12.5 fps, resulting in a PSNR of about 34 dB in the error-free case. Nine simulations are conducted for each GOB in the fifth encoded frame (dotted lines). The solid line represents the averaged result, indicating that a residual loss of approximately 1 dB still remains in the sequence after 3 seconds due to temporal error propagation.

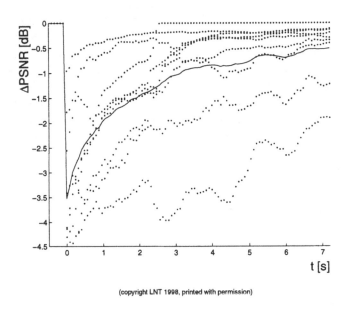

Figure 6.3   Decrease in PSNR after concealed loss of one GOB

6.4   MODELING OF TRANSMISSION ERROR EFFECTS

In the following subsections we derive a simple model for D_v, i.e., the distortion that is caused by transmission errors and propagates due to the recursive DPCM structure in the decoder. We are interested in the signal $v[x, y, t] = c[x, y, t] - r[x, y, t]$ and in particular in its variance $\sigma_v^2[t]$ after t decoded frames. Based on the model we will explain how spatial filtering and INTRA update affect $\sigma_v^2[t]$.

6.4.1 Two-Dimensional Case

First we restrict the model to the two-dimensional case with the discrete spatial and temporal variables x and t and extend the results to the more general case later on. As illustrated in Fig. 6.1 the decoder can be regarded as a linear system $H_t(\omega_x)$ with parameter t. The power spectral density (PSD) of $v_t[x]$ can then be written as

$$\Phi_{vv,t}(\omega_x) = |H_t(\omega_x)|^2 \Phi_{uu}(\omega_x), \tag{6.3}$$

and $\sigma_v^2[t]$ can be obtained simply as

$$\sigma_v^2[t] = \frac{1}{2\pi} \int\limits_{-\pi}^{+\pi} \Phi_{vv,t}(\omega_x) d\omega_x. \tag{6.4}$$

In hybrid video coding a spatial filter $F(\omega_x)$ is used in the prediction loop either explicitly as e.g. the loop filter in H.261 or implicitly for the interpolation of sub-pel positions as in H.263 and MPEG-2. For the two-dimensional case we assume that the same filter is applied in each time step. Then the impulse response $h_t[x]$ can be defined recursively as $h_t[x] = h_{t-1}[x] * f[x]$, where '$*$' denotes convolution and $f[x]$ is the impulse response of the filter. Based on the central limit theorem we expect $h_t[x]$ to be Gaussian for large t, because it is the result of many convolutions with $f[x]$. Therefore the magnitude of the transfer function of the decoder can be approximated in the base band $|\omega_x| < \pi$ by

$$|\hat{H}_t(\omega_x)| = \exp\left(-\frac{\omega_x^2 t \sigma_f^2}{2}\right), \tag{6.5}$$

where σ_f^2 is defined as

$$\sigma_f^2 = \sum_x x^2 f[x] - \left(\sum_x x f[x]\right)^2. \tag{6.6}$$

Note that this approximation is useful for any filter with $F(0) = 1$ and $f[x] \geq 0$ for all x. In the case of linear interpolation with

$$f[x] = \begin{cases} 1/2 & \text{for} \quad x = 0, 1 \\ 0 & \text{else,} \end{cases} \tag{6.7}$$

we obtain $\sigma_f^2 = 1/4$.

In addition to the Gaussian approximation for $H_t(\omega_x)$ we also approximate the PSD of the introduced error signal $u[x]$ by

$$\hat{\Phi}_{uu}(\omega_x) = \sigma_u^2 \sqrt{4\pi\sigma_g^2} \exp(-\omega_x^2\sigma_g^2), \qquad (6.8)$$

i.e., a Gaussian PSD with the same energy σ_u^2. The parameter σ_g^2 determines the shape of the PSD and can be used to match (6.8) with the true PSD. With the given approximations for $|H_t(\omega_x)|$ and $\Phi_{uu}(\omega_x)$ we can solve (6.4) directly yielding

$$\hat{\sigma}_v^2[t] = \sigma_u^2 \sqrt{\frac{\sigma_g^2}{\sigma_g^2 + t\sigma_f^2}} = \sigma_u^2\alpha[t], \qquad (6.9)$$

where $\alpha[t]$ is the power transfer factor after t time steps.

6.4.2 Extension to Three-Dimensional Case

In the following we focus on spatial filtering caused by motion compensated prediction with half-pel accuracy and bilinear interpolation. For the extension of the above results to the three-dimensional case with the discrete variables x, y, and t, it has to be considered that individual image regions undergo different filtering operations. For each macroblock a different motion vector is selected and depending on its sub-pel fractions $(sx, sy) \in \{0.0, 0.5\}$ either no filtering, only horizontal, only vertical, or horizontal and vertical (diagonal) spatial filtering is applied. The probabilities of using these filters are denoted p_N, p_H, p_V, and p_D respectively and are approximately $1/4$ for moving areas.

In the process of encoding, different combinations of filters are applied to an individual image region after t time steps. For example, a macroblock with the motion vectors $(0.0, 0.0)$ and $(0.0, 0.5)$ in two successive frames undergoes only one vertical filter operation, while a macroblock with motion vectors $(0.5, 0.5)$ and $(0.5, 0.5)$ undergoes two horizontal and two vertical filter operations. Based on (6.9) it can be shown that the variance after h horizontal and v vertical filter operations is

$$\sigma_v^2[h, v] = \sigma_u^2\alpha[h]\alpha[v]. \qquad (6.10)$$

The probability that an image region is filtered h times horizontally and v times vertically can be described by a two dimensional pdf $p_t[h, v]$ with parameter t. For $t = 0$ no filter operations have yet been performed, i.e., $p_0[0, 0] = 1$ and otherwise zero. For $t = 1$ the pdf is given by the above probabilities, i.e., $p_1[0, 0] = p_Z$, $p_1[1, 0] = p_H, p_1[0, 1] = p_V, p_1[1, 1] = p_D$, and otherwise zero. For $t > 1$ the pdf can be defined recursively according to (6.11) under the assumption that the sub-pel fractions are independent in each frame at a given location.

$$p_t[h, v] = p_{t-1}[h, v] * p_1[h, v] \tag{6.11}$$

With this assumption we can formulate the variance of the signal v for the three-dimensional case as

$$\hat{\sigma}_v^2[t] = \sigma_u^2 \sum_h \sum_v \alpha[h]\alpha[v]p_t[h, v]. \tag{6.12}$$

This equation does not consider INTRA coded macroblocks, which will cause a faster decrease in error energy. If the INTRA mode is selected randomly for n out of N macroblocks per frame, the effect on the variance can be modeled as an additional leakage $\beta = 1 - n/N$ resulting in

$$\hat{\sigma}_v^2[t] = \beta^t \sigma_u^2 \sum_h \sum_v \alpha[h]\alpha[v]p_t[h, v]. \tag{6.13}$$

Note that (6.10) - (6.13) are derived under the assumption of a Gaussian PSD for the error signal u.

6.4.3 Comparison with Simulation Results

To achieve a good match between the simulation results and the model we need to set the parameter σ_g^2 that describes the shape of the assumed PSD $\hat{\Phi}_{uu}$ (6.8). We found that a value of $\sigma_g^2 = 1.3$ yields a good correspondence between the theoretical results and our measurements. We use this value in the following.

Four simulations (A-D) are performed under different conditions with respect to spatial filtering and INTRA update. The general simulation conditions are identical to those in Fig. 6.3. All macroblocks are encoded in the INTER mode, i.e., no macroblocks are UNCODED or encoded in INTRA mode except for simulation D, where 9 out of 99 macroblocks per frame are randomly coded in INTRA mode. In simulation A the sub-pel fractions (sx, sy) are forced to $(0.5, 0.0)$ (i.e. $p_H = 1$), such that only horizontal filter operations are performed. Similarly, in simulation B the sub-pel fractions are forced to $(0.5, 0.5)$ (i.e. $p_D = 1$), such that only diagonal filter operations are performed. For the remaining simulations C and D, the sub-pel fractions are not altered but remained as selected by the TMN5 mode decision. The simulation results are shown in Fig. 6.4.

As can be seen, the model provides a good estimate of the error energy. Note that in simulation A the energy decreases proportional to $\sqrt{1/t}$, while it decreases proportional to $1/t$ in simulation B. Though the conditions for those two simulations are artificial, the results show how spatial filtering can effect error recovery. In the extreme case of $P_N = 1$, i.e. motion compensation with integer-pel accuracy, the error would remain indefinitely.

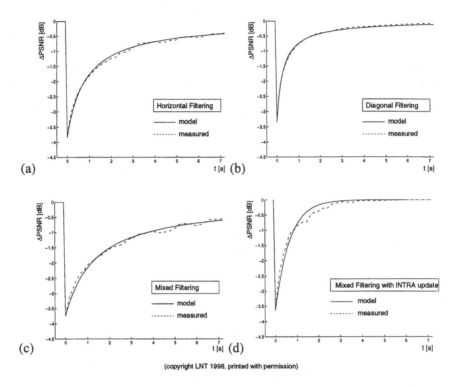

(copyright LNT 1998, printed with permission)

Figure 6.4 Comparison of measured loss of picture quality with model for (a) forced sub-pel fractions (0.5, 0.0), (b) forced sub-pel fractions (0.5, 0.5), (c) no forced sub-pel fractions, and (d) as (c) but with 9 out of 99 macroblocks per frame in INTRA mode.

Summarizing this section, it can be seen that spatial filtering and INTRA update can have significant influence on error recovery in hybrid video coding. Interestingly, spatial filtering can also improve coding performance. If, for example, the loop filter in H.261 is used during encoding, the gain in PSNR is up to 2 dB [24] while error recovery is also improved significantly. However, the leakage in the DPCM loop of standardized video codecs is not strong enough to be used for error robustness alone. Quick error recovery is only possible when INTRA coding is used as illustrated in simulation D. However, INTRA coding reduces coding performance significantly. Therefore it is necessary to restrict the use of this mode as much as possible, as intended by the channel adaptive approaches described in section 6.5.

6.5 ERROR COMPENSATION BASED ON A FEEDBACK CHANNEL

As shown in the previous section, the remaining distortion after error concealment of corrupted GOBs may remain visible in the image sequence for several seconds. The Error-Tracking approach utilizes the INTRA mode to stop temporal error propagation but limits its use to severely affected image regions only. During error-free transmission, the more effective INTER mode is utilized and the system therefore adapts effectively to varying channel conditions. Note that this approach requires that the encoder has knowledge of the location and extent of erroneous image regions at the decoder. As will be shown below, this can be achieved by utilizing a feedback channel.

The feedback channel is used to send Negative Acknowledgments (NAKs) back to the encoder. NAKs report the temporal and spatial location of GOBs that could not be decoded successfully and had to be concealed. For H.263, the temporal and spatial location of a GOB can be encoded by the time reference (TR, 8 bit) and group number (GN, 5 bit). The resulting rate on the feedback channel is then mainly determined by the GOB error rate, but will in general be very small.

Based on the information of a NAK, the encoder can reconstruct the resulting error distribution in the current frame. To do so, the encoder could store its own bitstream and simulate the loss of the reported GOB while decoding it again. Note that the encoder will have to use the same concealment strategy as the decoder when simulating the loss. While this approach is not feasible for a real-time implementation, it shows that the encoder itself "knows" all the information necessary for the reconstruction of spatio-temporal error propagation. For a practical system, the error distribution has to be estimated with a low complexity algorithm, which will be described in Section 6.6. For the rest of this section, however, we assume that the encoder gains complete knowledge of the true error distribution. Then, the coding control of the encoder can be modified to effectively stop error propagation by selecting the INTRA mode whenever a MB is severely distorted. On the other hand, if error concealment was successful and the error of a certain MB is only small, the encoder may decide that INTRA coding is not necessary.

Fig. 6.5 shows averaged simulation results for an assumed round-trip delay of 800 ms and identical simulation conditions as in Fig. 6.3. Compared to the case without error compensation, the picture quality recovers rapidly as soon as INTRA coded MBs are received. Note that the round trip delay is much greater than 250 ms, which is a common requirement for conversational services. It is also important to note that our approach does not add any additional delay. A longer delay just results in a later start for the error recovery. Considering the slow recovery for "Concealment only", NAKs may still be useful after several seconds.

Error-Tracking represents the approach described above including the reconstruction of spatio-temporal error propagation. *Same-GOB* is a simplified version, where the spatial error propagation is not taken into account and the entire reported GOB

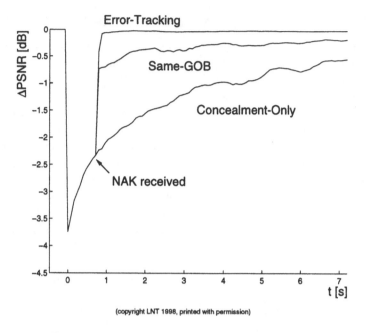

Figure 6.5   Error recovery with feedback channel

is INTRA refreshed. The Same-GOB strategy does not require the reconstruction of spatial error propagation and therefore exhibits lower complexity. However, it does not consider the propagation of errors across the borders of the GOB. Though the difference between the two strategies seems to be small in the average case (0.5 dB), annoying artifacts may be avoided by Error-Tracking for particularly unfavorable cases.

A perfect reconstruction of the error propagation as part of the coder control, while theoretically possible, would be computationally extremely demanding. In the next section we present a low complexity algorithm that estimates the spatio-temporal error propagation with a sufficient accuracy. This scheme is of particular interest to real-time implementations since its usage of memory and the processing load are very modest.

6.6   LOW COMPLEXITY ESTIMATION OF ERROR PROPAGATION

The spatio-temporal dependencies of MBs in successive frames arise from motion compensated prediction when coding MBs in INTER mode. These dependencies, together with an error severity measure, have to be stored at the encoder in order to provide enough information for rapid reconstruction of spatio-temporal error propagation. This can be achieved by the following algorithm.

Assume N macroblocks within each frame enumerated $mb = 1...N$ in transmission order from top-left to bottom-right. Let $\{n_{err}, mb_{first}, mb_{last}\}$ be the content of an NAK sent to the encoder, where $mb_{first} \leq mb \leq mb_{last}$ indicates a set of erroneous macroblocks in frame n_{err}. For the case that complete GOBs are discarded and concealed, as described in section 6.3, mb_{first} and mb_{last} would correspond to the first and last MB of lost GOBs.

To evaluate the NAK, the encoder must continuously record information during the encoding of each frame. First, the initial error "energy" $E_0(mb, n)$ that would be introduced by the loss of macroblock mb in frame n needs to stored. For the zero-MV error concealment strategy described in section 6.3.1, $E_0(mb, n)$ may be computed as the Summed Absolute Difference (SAD) of macroblock mb in frame n and n-1. Second, the number of pixels transferred from macroblock mb_{source} in frame $n - 1$ to macroblock mb_{dest} in frame n is stored in dependencies $d(mb_{source}, mb_{dest}, n)$. These dependencies can be derived from the motion vectors. Fig. 6.6 shows an example for the dependencies introduced by the motion vector (-4,-11) in MB 30. The prediction is formed from MB 18, 19, 29 and 30 using 44, 132, 20 and 60 pixels respectively. According to the above notation, $d(18, 30, n)$ would then for example be set to 44.

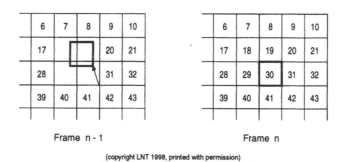

Frame n - 1   Frame n

Figure 6.6   Motion compensation of MB 30 with motion vector (-4,-11).

In practice, error tracking information will be stored in a cyclic data structure covering the last M frames. The value of M has to be chosen depending on the maximum round trip delay and the encoded frame rate. Because the typical frame rate for low bit-rate video is rather low (e.g. 10 fps), and the round trip delay for conversational services should be in the order of 250 ms, typical values for M will be less than 10. Note that the area used for motion compensation of a particular macroblock can only overlap with a maximum of 4 macroblocks. Therefore, only 4 dependencies will have to be stored for each macroblock, resulting in small storage requirements.

The evaluation of the stored error tracking information is carried out as follows. Assume that an NAK arrives before frame n_{next} is encoded, such that $n_{next} > n_{err}$.

Then, the estimated error $E(mb, n_{err})$ in macroblock mb and frame n_{err} is initialized as

$$E(mb, n_{err}) = \begin{cases} E_0(mb, n_{err}) & \text{for} \quad mb_{first} \leq mb \leq mb_{last} \\ 0 & \text{else} \end{cases} . \qquad (6.14)$$

For subsequent frames n, with $n_{err} < n < n_{next}$, the error may be estimated recursively as

$$E(mb, n) = \sum_{i=1}^{N} \frac{E(i, n-1)d(i, mb, n)}{256}, \qquad (6.15)$$

where a uniformly distributed error in each macroblock is assumed after each iteration. Note that the evaluation of (6.15) does not actually require N multiplications for each macroblock but a maximum of 4 (see above). Therefore, only $4N$ multiplications are required for the whole frame which results in a small computational complexity of the algorithm.

After the last iteration, the error distribution $E(mb, n_{next} - 1)$ is available to the encoder and may be incorporated into the mode decision of the next frame. In the following simulations, we encode the same number of macroblocks in INTRA mode for both channel adaptive schemes, i.e., Error-Tracking and Same-GOB. Therefore, we are using a budget that is increased by the number of erroneously reported MBs each time a NAK is received. Furthermore, we restrict the number of INTRA coded macroblocks per frame to 11 in order to avoid excessive fluctuations in picture quality while maintaining a constant frame rate. Therefore, in the case of Error-Tracking the 11 most affected MBs are encoded in INTRA mode in each frame, where $E(mb, n_{next} - 1)$ is used to perform the ranking. In the case of the Same-GOB strategy, the INTRA update of erroneous GOBs may also be distributed over several frames in order to comply with the restriction of a maximum number of 11 INTRA coded macroblocks per frame.

6.7 SIMULATION OF WIRELESS DECT CHANNEL

In this section, we investigate the performance of the proposed error compensation techniques in a mobile environment by simulating the transmission over a wireless DECT (Digital European Cordless Telephony) channer. Our simulations ae based on bit error sequences that are generated assuming Rayleigh fading and a y of 14 km/h. Carrier-to-Noise Ratios (E_b/N_0) in the range from 20 to 30 dB were investigated, with corresponding Bit Error Rates (BERs) as summarized in Table 6.1. The bit error sequences exhibit severe burst errors and provide a total bit-rate of 80 kbps, which is the available bit-rate in the double slot format of DECT.

Table 6.1   Summary of test channel parameters.

E_b/N_0 [dB]	Bit Error Rate	Block Error Rate
20	0.002578	0.017075
22	0.001646	0.011250
24	0.001025	0.007575
26	0.000644	0.004675
28	0.000390	0.002775
30	0.000234	0.001725

Forward Error Correction (FEC) is based on a BCH code table [25] from which a code with block size $n = 255$ bit, $k = 179$ information bits, and $t = 10$ correctable errors per block is selected. The resulting Block Error Rates (BERs) for this FEC are also summarized in Tab. 6.1. The same FEC is used for all approaches, i.e Random-Intra, Same-GOB, and Error-Tracking. If the errors within a block cannot be corrected by FEC, the block is declared erroneous and all affected GOBs are discarded by the video decoder which uses the described error concealment afterwards. As discussed in section 6.3, we match the distortion caused by coding (D_c) for all approaches. Because the video bit-rate is identical for all approaches, the coding distortion is mainly determined by the number of INTRA coded MBs. For the channel adaptive approaches, this number is driven by the quality of the channel, resulting in an increased quality for increasing E_b/N_0. For the random INTRA case, the average number of INTRA coded MBs per frame was matched for each channel. Therefore the total amount of INTRA coded MBs in each simulation is about the same, however, the different approaches will use this budget more or less efficiently.

To complete the description of the transmission system, we note that the feedback channel is assumed to be error-free at a constant delay of 100 ms. For the effective delay that describes the delay from the encoding of a frame to the reception of a NAK, the delay caused by buffering at the encoder and decoder has to be considered as well. The effective delay was in the order of about 300 ms.

The simulations are performed with the test sequence *Mother and Daughter* for the zero-MV and true-MV error concealment. The first 300 frames are coded at a frame rate of 12.5 fps. 30 simulations are averaged for different offsets in the bit error sequence, such that each simulation starts two seconds after the preceding one in order to distribute the burst errors at varying temporal and spatial locations in the sequence. The sequence is encoded at a fixed frame rate with only little variations in the number of bits per frame to avoid frame jitter. Therefore an increased number of INTRA coded

192

MBs per frame directly influences the picture quality in that frame. To avoid excessive fluctuations in picture quality, the maximum number of INTRA coded MBs per frame were restricted to 11 for the SG and ET approaches.

Figs. 6.8 and 6.7 show results of the picture quality after coding (D_c), the picture quality after transmission (D_r) and the loss of picture quality caused by transmission errors (D_v). The averaged values over all frames and all offsets into the bit error sequence are plotted over E_b/N_0. Both feedback based approaches perform significantly better than the random update scheme. At $E_b/N_0 = 20$ dB the gain is about 0.5 dB for the true-MV concealment and about 1.0 dB for the zero-MV concealment. The Error-Tracking strategy always outperforms the Same-GOB strategy. However, the PSNR gain is only marginal in the average case and decreases towards better channel conditions, i.e., higher E_b/N_0. The average loss of picture quality can be improved significantly by advanced error concealment. For the upper bound of true-MV error concealment, the Random Intra scheme performs almost as good as any of the feedback based schemes with the simple zero-MV error concealment. However, when the INTRA update schemes are compared with the same error concealment then the same ranking in performance can be observed.

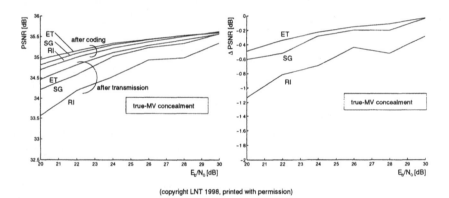

(copyright LNT 1998, printed with permission)

Figure 6.7 Picture quality after coding and transmission for the true-MV concealment. The INTRA update techniques Error-Tracking, Same-GOB, and Random-Intra are denoted ET, SG, and RI respectively. The test sequence is Mother and Daughter.

Note that for individual error events the Error Tracking strategy can provide a significant advantage that cannot be reflected appropriately by averaged distortion measures. This is illustrated in Fig. 6.9, that shows the loss of picture quality over time for an individual simulation. For the first two burst errors, the Error Tracking and Same-GOB strategy perform very similar. The third burst error, however, causes spatial error propagation resulting from the particular motion in this part of the sequence. As

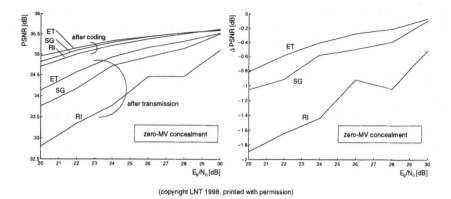

(copyright LNT 1998, printed with permission)

Figure 6.8 Picture quality after coding and transmission for the zero-MV concealment. The INTRA update techniques Error-Tracking, Same-GOB, and Random-Intra
are denoted ET, SG, and RI respectively. The test sequence is Mother and Daughter.

can be seen, the Same-GOB strategy cannot fully compensate this error event and a
loss of picture quality of about 1 dB remains.

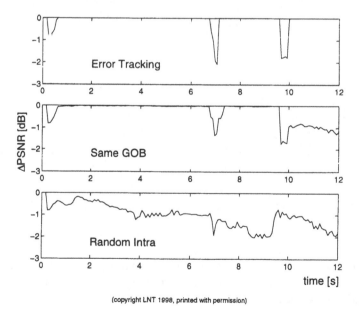

(copyright LNT 1998, printed with permission)

Figure 6.9 Loss of picture quality for an individual simulation (E_b/N_0 =30 dB, Mother
and Daughter).

194

To further illustrate the advantage of the Error Tracking strategy, example frames from the *Foreman* test sequence after the loss of 2 GOBs are presented in Fig. 6.10. Because the error in frame 75 has propagated from its original location, the Same-GOB strategy cannot remove it successfully in frame 90. Only when Error-Tracking is employed, the propagation can be stopped entirely. Because these error events are relatively seldom and the resulting errors are concentrated in a small part of the image, they do not significantly affect average PSNR values. The subjective image quality, however, can be severely affected. Therefore, Error Tracking provides increased robustness for particularly unfavorable cases, that may otherwise cause annoying artifacts.

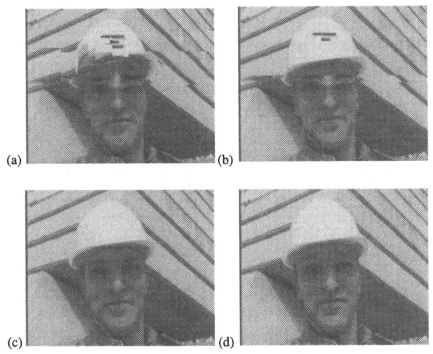

(a)   (b)

(c)   (d)

(copyright LNT 1998, printed with permission)

Figure 6.10   (a) Frame 90 of sequence Foreman after two GOBs were lost and concealed in frame 75 (b) Same-GOB strategy (c) Error-Tracking (low complexity algorithm) (d) Frame 90 without GOB loss in frame 75

6.8 CONCLUSIONS

Spatio-temporal error propagation is an important problem in robust video transmission because successive frames can be affected by a single transmission error. In this chapter we investigated how spatial filtering and random INTRA updating influence the propagation of error energy and proposed a model that explains experimental results from an H.263 decoder. For the performed simulations the model predicts the measured behavior fairly accurately.

We then compared three different INTRA update schemes that can be used to restrict error propagation in the H.263 video compression standard. The comparison is based on a simulated transmission over a wireless DECT channel at various values for E_b/N_0. The Same-GOB and Error-Tracking strategy are both based on a feedback channel and show a significant gain over the non-adaptive, Random-Intra update scheme. Within the feedback based schemes, the Error-Tracking strategy shows a consistent gain over the Same-GOB strategy that decreases at higher E_b/N_0. Though the average gain is less than 0.3 dB, annoying artifacts can be avoided in particular unfavorable cases by considering spatial error propagation. Because these error events are relatively seldom and the resulting errors are concentrated in a small part of the image, they do not significantly affect average PSNR values. The subjective image quality, however, can be severely affected. The low complexity algorithm proposed for the reconstruction of spatio-temporal error propagation approximates the actual error propagation and is of particular interest to real-time implementations of the Error-Tracking strategy. It is important to note that no additional delay is introduced in the feedback based systems, since the round-trip delay only determines the time until error recovery.

Error concealment has a significant influence on the overall performance. For the upper bound of error concealment with the true motion vector, the Random-Intra scheme performs almost as good as any of the feedback based schemes when those are used with a zero motion vector concealment. For the same error concealment, however, the Error-Tracking strategy is still advantageous.

The usage of a feedback channel for error robust video transmission is recommended by the ITU-T Study Group 16 as an integral part of the mobile extensions of H.324. The low complexity algorithm for reconstruction of spatio-temporal error propagation described in this chapter forms an informative appendix in H.263 (Appendix II). For convenient use in a H.324 terminal, the H.245 control standard has been extended to include the feedback message "videoNotDecodedMBs", that can be used to indicate the loss of macroblocks.

References

[1] ITU-T Recommendation H.263, "Video Coding for Low Bitrate Communication," 1996.

196

[2] Y. Wang and Q.-F. Zhu, "Error Control and Concealment for Video Communication: A Review", Proceedings of the IEEE, vol. 86, no. 5, pp. 974-997, May 1998.

[3] LBC Doc. LBC-95-304 (ITU-T Study Group 15, Working Party 15/1), "Draft Recommendation AV.26M," Robert Bosch GmbH, Darmstadt, 1995.

[4] R. Fischer, P. Mangold, R.M. Pelz, G. Nitsche, "Combined Source and Channel Coding for Very Low Bitrate Mobile Visual Communication Systems," in Proc. Int. Picture Coding Symposium, pp. 231-236, 1996.

[5] K. Illgner and D. Lappe, "Mobile multimedia communications in a universal telecommunications network," in Proc. of SPIE Conf. on Visual Communications and Image Processing, pp. 1034-1043, 1995.

[6] B. Girod, U. Horn, and B. Belzer, "Scalable video coding with multiscale motion compensation and unequal error protection," in Multimedia Communications and Video Coding, Y. Wang, S. Panwar, S.P. Kim, and H.L. Bertoni (eds.), New York: Plenum Press, 1996, pp. 475-482.

[7] R. Mann Pelz, "An unequal error protected px8 kbit/s video transmission for dect," in Vehicular Technology Conference, pp. 1020-1024, 1994.

[8] LBC Doc. LBC-95-309 (ITU-T Study Group 15, Working Party 15/1), "Sub-videos with retransmission and intra-refreshing in mobile/wireless environments," National Semiconductor Corporation, Darmstadt, 1995.

[9] LBC Doc. LBC-96-033 (ITU-T Study Group 15, Working Party 15/1), "An error resilience method based on back channel signalling and FEC," Telenor Research, San Jose, 1996.

[10] N. Färber, E. Steinbach, B. Girod, "Robust H.263 Compatible Video Transmission Over Wireless Channels," in Proc. Int. Picture Coding Symposium, pp. 575-578, 1996.

[11] E. Steinbach, N.Färber, and B. Girod, "Standard Compatible Extension of H.263 for Robust Video Transmission in Mobile Environments," IEEE Trans. Circuits and Sys. for Video Tech., vol. 7, no. 6, pp. 872-881, Dec. 1997

[12] B. Girod, N. Färber, E. Steinbach, "Error-Resilient Coding for H.263," in Insights into Mobile Multimedia Communication, D. Bull, N. Canagarajah, A. Nix (eds), Academic Press, To be published.

[13] S. Lin, D.J. Costello, and M.J. Miller, "Automatic repeat error control schemes," IEEE Communications Magazine, vol. 22, pp. 5-17, 1984.

[14] A. Heron, N. MacDonald, "Video Transmission over a radio link using H.261 and DECT," IEE conference publications, no. 354, pp. 621-624, 1992.

[15] M. Khansari, A. Jalali, E. Dubois, and P. Mermelstein, "Low Bit-Rate Video Transmission over Fading Channels for Wireless Microcellular Systems," IEEE

Transactions on Circuits and Systems for Video Technology, vol. 6, no. 1, pp. 1-11, Feb. 1996.

[16] C. Chen, "Error Detection and Concealment with an Unsupervised MPEG2 Video Decoder," Journal of Visual Communication and Image Representation, Vol. 6, No. 3, pp. 265-278, Sep. 1995.

[17] P. Haskell, and D. Messerschmitt, "Resynchronization of motion compensated video affected by ATM cell loss," in Proc. ICASSP, vol.3, pp. 545-548, 1992.

[18] M. Wada, "Selective recovery of video packet loss using error concealment," IEEE Journal on Selected Areas in Communications, vol. 7, pp. 807-814, 1989.

[19] K. Tzou, "Post Filtering for Cell Loss Concealment in Packet Video," SPIE Visual Communications and Image Processing IV, vol. 1199, pp. 1620-1628, 1989.

[20] W.-M. Lam, A.R. Reibman, and B. Lin, "Recovery of lost or erroneously received motion vectors," in Proc. ICASSP, vol. 5, 1993.

[21] G. Wen and J. Villasenor, "A class of reversible variable length codes for robust image and video coding," in IEEE Int. Conf. Image Proc., vol. 2, pp. 65-68, Santa Barbara, CA, Oct. 1997.

[22] R. Talluri, "Error-Resilient Video Coding in the ISO MPEG-4 Standard," IEEE Communications Magazine, vol. 36, no. 6, pp. 112-119, June 1998.

[23] LBC Doc. LBC-96-186 (ITU-T Study Group 15, Working Party 15/1), "Definition of an error concealment model (TCON)," Telenor Research, Boston, 1995.

[24] B. Girod, N. Färber, E. Steinbach, "Performance of the H.263 Video Compression Standard", Journal of VLSI Signal Processing: Systems for Signal, Image, and Video Technology, vol. 17, pp. 101-111, Nov. 1997.

[25] G.C. Clark, Jr., and J.B. Cain, Error-Correction Coding for Digital Communications, New York: Plenum Press, 1988.

7 ERROR CONCEALMENT IN ENCODED VIDEO STREAMS

Paul Salama, Ness B. Shroff, and Edward J. Delp

Video and Image Processing Laboratory (*VIPER*)
School of Electrical and Computer Engineering
Purdue University
West Lafayette, IN
{salama,shroff,ace}@ecn.purdue.edu

Abstract: In ATM networks cell loss causes data to be dropped, which results in the loss of entire macroblocks when MPEG video is being transmitted. In order to reconstruct the missing data, the location of these macroblocks must be known. We describe a technique for packing ATM cells with compressed data, with the aim of detecting the location of missing macroblocks in the encoded video stream. This technique also permits proper decoding of correctly received macroblocks, and thus prevents the loss of ATM cells from affecting the decoding process. We also describe spatial and temporal techniques for the recovery of lost macroblocks. The spatial techniques fall into two categories: deterministic and statistical. A deterministic spatial approach we provide aims at reconstructing each lost pixel by spatial interpolation from the nearest undamaged pixels. Another, recovers lost macroblocks by minimizing inter-sample variations within each block and across its boundaries. In the statistical approach, each frame is modeled as a Markov Random Field, and a maximum a posteriori (MAP) estimate of the missing macroblocks is obtained based on this model. The MAP estimate for each pixel within a lost macroblock is obtained by means of the iterative conditional modes (ICM) algorithm. The iterative method is guaranteed to converge to a global maximum, even though the global maximum is not unique. It is shown that, for each pixel, the median of its neighbors is a MAP estimate. In temporal reconstruction, a search is carried out over a reference frame for the macroblock sized region that will maximize the posterior distribution of the lost macroblock given its neighbors.

199

200

Acknowledgments

This work was supported by a grant from the AT&T Foundation and partially supported by the National Science Foundation under Grant No. NCR-9624525.

7.1 INTRODUCTION

Video has traditionally been recorded, stored and transmitted in analog form. However, digital video has rapidly gained popularity [1, 2, 3, 4]. Advances in digital video technology and storage have made it possible to integrate digital video into a number of multimedia applications. In response to a growing demand for a common format for coding and storing digital video, the International Organization for Standardization (ISO) established the Moving Pictures Expert Group (MPEG) to develop standards for coded representations of moving pictures and associated audio, for storage on digital media, cable television distribution, and digital terrestrial television, among other applications. MPEG, formally known as group ISO/IEC JTC 1/SC 29/WG 11, developed the ISO/IEC 1172 and ISO/IEC 13818 standards [5, 6], commonly known as the MPEG–1 and MPEG–2 standards, to meet the above mentioned needs.

The main objective of the MPEG–1 standard was to compress 4:1:1 CIF digital video sequences (31.1 Mbits/s uncompressed) to a target bit rate of 1.5 Mbits/s. This required a 30:1 compression ratio, which was to be achieved while providing such features as random access, fast forward, or fast and normal reverse playback [7]. The standard defines a generic decoder but leaves the implementation of the encoder open to individual designs.

MPEG–2, the result of the second phase of the work done by MPEG, was originally intended to be applicable to a wide class of applications, as well as support a compressed bit rate of 5 Mbits/s for 4:2:2 CCIR601 digital video. Among the original requirements were compatibility with MPEG–1, excellent picture quality, flexibility of input format, random access capability, fast forward capability, reverse play capability, resilience to bit errors, and bit stream scalability [8, 6, 9, 10].

Currently the MPEG is developing two more standards, MPEG–4 and MPEG–7. MPEG–4 will provide a set of technologies to satisfy the needs of authors, service providers, and end users alike. It will enable the production of content that has far greater re-usability, has greater flexibility than is possible today with individual technologies such as digital television, animated graphics, World Wide Web (WWW) pages and their extensions. It will also offer a consistent interface to indicate the quality of service required by the individual media, as well as allow higher levels of interaction with content, within the limits set by the author [11]. MPEG–7, on the other hand is intended to provide complementary functionality to the other MPEG standards. It will be a standardized description of various types of multimedia information. This description will be associated with the content, to allow fast and efficient searching

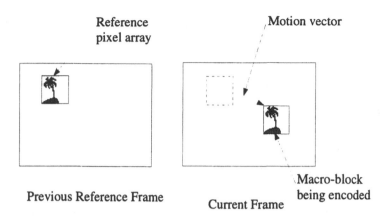

Reference pixel array

Motion vector

Previous Reference Frame

Current Frame

Macro-block being encoded

Figure 7.1 Motion compensated prediction using past frames. Macroblocks in the current frame are compared to macroblock sized regions in the previous frame for a close match up.

for material that is of interest to the user. MPEG-7 is formally known as Multimedia Content Description Interface [12].

Digital video is compressed by reducing the temporal and spatial redundancies in every frame. The MPEG-1 and MPEG-2 standards define three types of frames, depending on the techniques utilized to compress them. They are the *intracoded* (I), *predicted* (P), and *bidirectionally-predicted* (B) Pictures. I pictures are compressed by reducing their spatial redundancies, while P and B pictures are compressed by reducing their temporal redundancies. P pictures are encoded using motion compensated prediction from a past I or P picture, as shown in Figure 7.1. B pictures, however, require both past and future reference frames (I or P pictures) for motion compensation, as illustrated in Figure 7.2. The outcome is that I pictures have the highest data rate and lowest motion artifacts, whereas B pictures have the lowest data rate and highest motion artifacts. Typical data rates are 1 bit per pixel, 0.1 bits per pixel, and 0.015 bits per pixel for I, P, and B pictures respectively.

Each video frame is divided into non-overlapping regions of pixels, known as macroblocks, each of which contains 16×16 pixels of the luminance component, 8×8 pixels of the C_b chrominance component and 8×8 pixels of the C_r chrominance component. An 8×8 pixel array is known as a block. Thus a macroblock consists of 4 luminance blocks and 2 chrominance blocks (one C_b block and one C_r block). An integral number of macroblocks, arranged in lexicographic order, are grouped together to from a slice. Slices within a picture can be of different sizes. The division of one picture into slices need not be the same as the division of any other picture. A slice can begin and end at any macroblock in a picture provided that the first slice begin at the

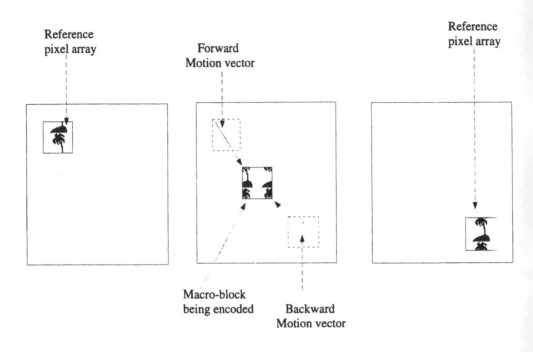

Figure 7.2 Motion compensated interpolation using past and/or future frames. A macroblock in the current frame is compared to macroblock sized regions in a previous frame and a future frame. The average of both matching regions is obtained. The displacements between the current macroblock and the matching regions in the previous and future frames are the forward and backward motion vectors, respectively.

top left of the picture, and the end of the last slice be the bottom right macroblock of the picture. There can be no gap between slices, nor can slices overlap. The minimum number of slices in a picture is one, and the maximum number is equal to the number of macroblocks. Several consecutive pictures, arranged in display order, are combined to form a structure known as a group of pictures (GOP). A GOP must contain at least one I picture, which may be followed by any number of I and P pictures. Any number of B pictures may interspersed between each pair of I or P pictures, and may also precede the first I picture. Since decoding B pictures requires both past and future pictures, the ordering of the pictures is changed prior to the transmission or storage of the encoded bit stream. The reference I and P pictures are placed before their dependent B pictures. GOPs facilitate the implementation of features such as random access, fast forward, or fast and normal reverse playback [7].

To generate I pictures, the two dimensional discrete cosine transform (DCT) [13] of each 8 × 8 block is obtained. This is done for all three color components. Each array of 8 × 8 DCT coefficients is then quantized, re arranged in a zig zag order, and then variable length coded [14].

At the receiving end the decoder reverses the zig zag scan, decodes the coded coefficient values, multiplies the coefficients by the quantizer step sizes, and performs the Inverse DCT (IDCT) [13] to obtain the pixel values. It is to be noted that due to quantization the resulting pixel values will differ from the original values.

Motion compensation is used to reduce temporal redundancy. This step is the most computationally intensive step in MPEG encoding. Motion compensated techniques are either predictive or interpolative. Motion compensated prediction assumes that a macroblock in the current frame or picture can be modeled as the translation of a macroblock sized region in a reference picture at some previous or future time, as shown in Figure 7.3. A search range is described, and a search conducted for that region that best matches the current macroblock. A match is found when a certain criterion or cost function, such as minimum mean square error (MMSE) or minimum mean absolute difference (MMAD), is minimized. The displacement between the centers of the macroblock and the best matching region in the reference frame is the motion vector [15]. When performing the matching process, macroblocks can be displaced by either integer or half pixel displacements. The choice of the matching criterion is left open to the individual designs.

To code P and B pictures, the encoder performs motion compensation for each macroblock. In the case of P pictures, past I or P frames are used as reference frames, whereas macroblocks in B frames are compared to macroblocks in past as well as future I or P frames [7]. If no match can be found for a particular macroblock, or if the minimized criteria is greater than a certain threshold, then that macroblock is intracoded. If however, the minimum value is less than the first threshold but greater than a second threshold, the motion vector is variable length coded and the difference

Reference
pixel array

Motion vector

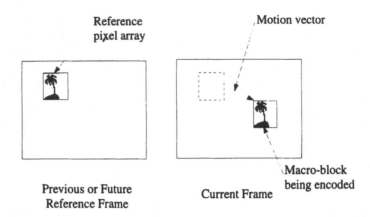

Previous or Future
Reference Frame

Current Frame

Macro-block
being encoded

Figure 7.3   Motion compensated prediction using past or future frames. A macroblock in the current frame is compared to macroblock sized regions in a previous frame and a future frame. The displacements between the current macroblock and the matching regions in the previous and future frames are the forward and backward motion vectors, respectively. The motion vector pointing to the most closely matching region is coded.

between the match and the original macroblock is intracoded. If the minimum value is less than both thresholds, then only the motion vector is variable length coded.

In addition to the coded motion vectors and DCT coefficients, an MPEG video bitstream contains macroblock addressing data, macroblock type, optional quantizer scale, 32 bit codewords used for synchronization purposes as well necessary header information required for the proper decoding of the video bitstream. Any damage incurred to the 32 bit codewords or the header data will inhibit the decoding process.

When delivering MPEG compressed video and audio bitstreams to a user, certain information such as presentation times or decoding times of pictures and audio frames have to be supplied. This is necessary for synchronized play back. Such information is not included within the compressed audio or compressed video bitstreams, but within what is known as the MPEG System Layer [16, 9, 17, 18].

The MPEG-1 and MPEG-2 system layer standards, address the problem of combining one or more data streams, whether they are compressed audio, compressed video or private data, into a single stream. The standards define data fields and specify semantic constraints on the data streams to enable decoders and/or encoders perform such functions as synchronized presentation of decoded information, the construction of a multiplexed stream, the management of buffers for coded data, and absolute time identification. The standards, however, do not specify the architecture or implementation of encoders or decoders.

There are two types of MPEG-2 system layer streams, Program Streams and Transport Streams [17, 18]. Program Streams were designed for work with digital storage media or channels that have minimal errors. They can only consist of elementary streams, such as compressed video and audio sequences. Transport Streams, however, are tailored for use in environments in which significant errors may occur. A Transport Stream may be constructed from elementary streams, from Program Streams, or from other Transport Streams. MPEG-1 system layer streams conform in syntax to Program Stream syntax, and can be decoded by Program Stream decoders.

When creating Program or Transport Streams, the streams to be multiplexed are segmented into packets. Each packet contains a header which includes such information as presentation time stamps, decoding time stamps, system clock reference fields, optional encryption, packet priority levels, trick mode indications that assist fast forward and slow motion, copyright protection, packet identifiers that identify the type of the data being carried, and packet sequence numbering.

Concurrent with the progress in multimedia signal processing, there has been a great interest in the networking community for the development of protocols that will harness the bandwidth offered by optical communication technologies. One such protocol is the Asynchronous transfer mode (ATM) [19, 20, 21, 22]. ATM has been chosen as the target protocol for broadband integrated services digital network (B-ISDN) because it offers a flexible transfer capability for a wide variety of data. These include video sequences, still images, and audio. In an ATM network, data is buffered as it arrives, segmented into blocks of fixed length, and inserted into what is known as an ATM cell. Cells are then transmitted across the network to the intended destination. Cells from different sources are multiplexed onto the channel thereby increasing the efficiency of the network. A cell, as shown in Figure 7.4, is 53 bytes in length and consists of an information field carrying user information (cell payload), and a header containing information used for routing and error detection purposes [20]. ATM networks can operate in either a connection oriented mode, in which cells are transmitted over designated paths, or a connectionless mode [19, 20]. Due to the nature of ATM networks, cells may be dropped because of network congestion, thus impacting the quality of the transmitted video. In this chapter we will discuss several techniques that can be implemented to repair damaged video sequences that have been impaired as a result of ATM cell loss.

7.2 CELL PACKING

In an MPEG sequence, consisting mainly of P and B frames, the sizes of the macroblocks will be smaller than the size of the user payload in an ATM cell. This is a consequence of the fact that coding most macroblocks in P or B pictures requires the coding of motion vectors and prediction errors, rather than the DCT coefficients of actual pixel values. This can be seen from Figures 7.5 and 7.6 respectively, which

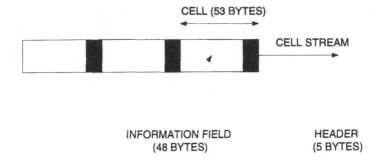

CELL (53 BYTES)

CELL STREAM

INFORMATION FIELD
(48 BYTES)

HEADER
(5 BYTES)

Figure 7.4   The ATM cell. Each cell consists of a 5 byte header and a 48 byte information field.

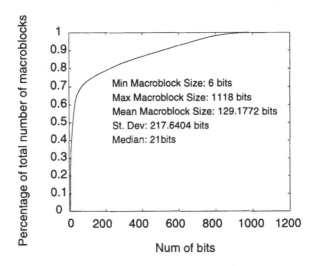

Figure 7.5   Cumulative histogram of the sizes, in bits, of the macroblocks belonging to the compressed version of the *flowergarden* sequence.

consist of cumulative histograms of the sizes, in bits, of the macroblocks belonging to the *flowergarden* and *football* sequences compressed using MPEG–1 to data rates of 5.05 Mbits/s and 3.81 Mbits/s, respectively.

In the event that cells are lost, the relative addressing of macroblocks is destroyed. Furthermore, pertinent information such as motion vectors, DCT coefficients, and user defined quantization scale is also destroyed. In this case, when attempting to reconstruct the macroblock, the decoder will use data belonging to other macroblocks.

In order to minimize the effect of cell loss, the transmitter could attempt to segment the incoming data in such a way that no macroblock data is split across cell boundaries.

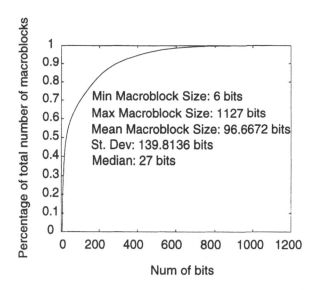

Figure 7.6   Cumulative distribution of the sizes, in bits, of the macroblocks belonging to the compressed version of the *football* sequence.

A macroblock greater in size than the remaining data in a cell, is placed in the next cell and the remaining space in the cell padded with zeros. Such a technique would be very efficient if the sizes of all the macroblocks in the sequence were much smaller than 48 bytes, the user payload in an ATM cell. Actual MPEG sequences, however, contain macroblocks of various sizes, including those whose sizes are greater than the size of an ATM cell user payload. Such a packing scheme would be extremely inefficient as shown in Figures 7.7 and 7.8 which illustrate the ensuing increases in data rates when such a packing scheme is employed to pack the compressed *flowergarden* and *football* sequences respectively, into ATM cells with different user payload sizes.

To avoid such excessive increase in data rates we present a new packing scheme. Our approach to packing cells is to pack important header information, such as packet headers as well as the header data necessary for the proper decoding of the compressed sequences, into high priority cells. All other data, such as motion vectors and DCT coefficients, are packed into low priority cells. An extra 9 bits, as shown in Figure 7.9, is then inserted at the start of each cell to provide the location of the first macroblock being packed into the cell. These extra bits are also used to indicate when a macroblock spans across more than one cell. They are then followed by another 7 bits that are used to provide the address of the macroblock relative to the slice to which it belongs. This is necessary since any macroblock is addressed relative to the preceding macroblock. The extra 16 bits are used to localize the loss of macroblocks within a frame while

208

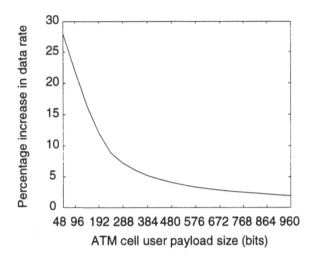

Figure 7.7   Percentage increase in data rate when the compressed *flowergarden* sequence is packed into ATM cells.   An integer number of macroblocks is packed into each cell, and extra spaces are filled in with zeros.

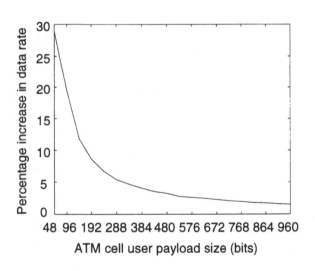

Figure 7.8   Percentage increase in data rate when the compressed *football* sequence is packed into ATM cells. The macroblocks are packed in such a way that no macroblock is split between two cells, and any extra space is padded with zeros.

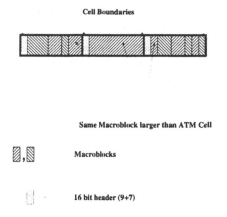

Figure 7.9 Packing scheme. Extra 9 bits are inserted at the start of each cell to indicate the location of the first macroblock being packed into the cell. These extra bits are also used to indicate when a macroblock spans across more than one cell. Another 7 bits are appended to the 9 bits, and used to provide the relative address of the macroblock with respect to the slice in which it is located.

maintaining the ease of decoding the correctly received macroblocks. This is crucial to MPEG streams, since a loss of a few bits can perturb the decoding process and result in the loss of entire frames. At the receiving end, the MPEG decoder will have to interpret the 16 bit header to determine the missing macroblocks as well as properly decode the rest of the sequence.

Several sequences were packed according to the new packing scheme. The overall increase in the data rate of every sequence was measured and found to be less than 5% of the original data rate. It is to be noted that 16 is the minimum number of bits necessary to provide the required information when CCIR601 images are coded such that each slice consists of an entire row of macroblocks.

7.3 PREVIOUS WORK IN CELL LOSS CONCEALMENT

A major issue that needs to be addressed when transmitting coded video over ATM networks is cell loss. Cell loss greatly offsets the decoder's synchronization, and by the time the decoder has resynchronized itself, major portions of a picture may have been lost. Hence, techniques for circumventing cell loss have to be developed. These methods must be capable of detecting where the lost data is located, and post processing the the damaged areas for error concealment.

Two methods for ameliorating the effect of ATM cell loss were proposed in [23]. It is assumed that both encoding and decoding are occurring simultaneously. It is also assumed once the decoder has detected cell loss, it communicates to the encoder the location of the damaged picture regions. The encoder then decides to do either of two things. The encoder may either disregard the affected areas as it continues to encode the video sequence, or it may reconstruct the affected areas and use them in the encoding process. Reconstruction is performed by replacing the damaged area by its corresponding area in the previous picture. This technique however introduces unacceptable delays, as the encoder will have to wait for the decoder to provide the location of damaged areas.

An alternative approach to ATM cell loss concealment is described in [24, 25, 26, 27, 28, 29, 30]. Encoded video data is segmented into low priority data such as high frequency DCT coefficients, and high priority data such as addresses of blocks, motion vectors, and low frequency DCT coefficients. Such prioritization is performed since motion vectors are crucial to the reconstruction of motion predicted regions. Likewise, most of the information for reconstructing blocks of data are available in the low frequency DCT coefficients. Prioritization is achieved by setting the priority bit to zero, for every cell carrying information that is deemed to be highly crucial to the proper reconstruction of the decompressed video sequence. In the event of network congestion, ATM cells carrying low priority data are discarded while those carrying high priority data are retained.

In [29] the transmission of spatially scalable MPEG2 video over ATM networks is discussed. Video is coded at different resolutions, with the lowest resolution being sent separately in high priority cells and the other resolution streams packed into low priority cells. At the decoder the low resolution frames are first retrieved and reconstructed. To attain higher resolutions, the decoded low resolution frames are then subsequently interpolated.

An additional technique to reduce the effect of cell loss was proposed in [27]. It is based on the fact that the data rate of a video sequence can be controlled by the quantization scale q, and by N, the number of intraframes in a sequence. Increasing q results in a decrease in the bit rate. Similarly, decreasing the number of intracoded frames will decrease the data rate. If the cell loss rate temporarily increases, a counter measure would then be to decrease N and increase q. This results in a smaller data rate and hence a smaller cell loss rate.

Alternative techniques to packetization that rely on interleaving of data are proposed in [28, 31, 32, 33]. The underlying idea behind either schemes in [31, 32] is to pack data, such as adjacent blocks or DCT coefficients from the same block into different cells to avoid a complete loss of whole blocks in the event that cell loss occurs. In [28] interleaving is performed at the slice level. It was found that packing every other slice into consecutive cells was sufficient to localize the effect of cell loss, at the mean cell loss rate of 10^{-3}, to a slice, and its neighboring slices kept intact. In [33], each frame is split into four bands and the data for each band packed together. This distributes the error due to cell loss over the entire frame, rather than localize it.

Post processing techniques for the sake of error concealment are addressed in [34, 35, 32, 36, 37, 38, 39, 40, 41]. In [34, 35, 32] it is assumed that partial loss of DCT coefficients has occurred. High frequency coefficients that have been lost are replaced by zeros. The remaining coefficients are estimated, by minimizing a cost function. The cost function is the sum, over all pixels in a block, of the square of the differences between adjacent coefficients. The difference between each coefficient and its neighbors is weighted by either zero or one, depending on whether or not that difference is to be included in the cost function. The method described in [36] was proposed for JPEG images but can be used for MPEG sequences as well. To perform error concealment, histograms of the DCT coefficients at locations (0,0), (0,1), (1,0), and (1,1) in intact regions, are compiled. These are then analyzed for correlations. The coefficients belonging to the missing macroblock are then estimated from those of the neighboring macroblocks. Instead of using all neighboring coefficients, only those that are believed to be correlated, based upon the histograms, are used in the reconstruction process. Using quadratic and linear polynomials to interpolate missing pixel values have also been investigated in [36]. An approach that estimates missing edges in each block from edges in the surrounding blocks was proposed in [38]. For each direction of an estimated edge, a version of the lost block is reconstructed by performing a number of one dimensional parallel to the edge. A final version of the missing block is obtained by merging all versions. In [37, 42, 43, 44, 45], lost macroblocks are constructed by temporal replacement or spatial interpolation. The decision to use a particular method is based on a measure of image activity. If there is no predominant motion within a frame then temporal replacement is used, otherwise the macroblock is reconstructed by means of spatial interpolation. The method of projections onto convex sets ([46]) for error concealment is described in [39]. To reconstruct a missing macroblock, it is first determined whether an edge actually a passes through it. This is achieved by applying the Sobel operator ([13]) to the surrounding macroblocks. A larger block of pixels is then constructed from initial pixel values of the lost macroblock and those of its neighbors. The Discrete Fourier Transform of the larger block is then obtained, and all coefficients lying outside a certain band that is oriented perpendicular to the edge direction are set to zero. The inverse transform is then obtained and those pixel values belonging to intact macroblocks are reset to their original values. The process is then repeated. Since the constraints imposed are convex sets, the process is guaranteed to converge.

In [28] a scheme that utilizes a hybrid of temporal and spatial interpolation is described. Again the decision to use spatial, or temporal, or both temporal and spatial interpolation is based on the amount of motion within a frame.

7.4 ERROR CONCEALMENT

The goal of error concealment is to detect the location of missing macroblocks

The goal of error concealment is to estimate missing macroblocks in MPEG data. The underlying idea is that there are still enough redundancies in the sequence to be exploited. In particular, in I frames it is possible to have a lost macroblock surrounded by intact macroblocks that are used to interpolate the missing data. In P and B pictures, it is possible to have entire rows of macroblocks missing. In this case spatial interpolation will not yield acceptable reconstructions. However, the motion vectors of the surrounding regions can be used to estimate the lost vectors, and the damaged region reconstructed via temporal interpolation [41]. Error concealment algorithms will have to be integrated into the decoder hardware, and should be capable of locating the missing macroblocks by interpreting the redundant data packed into ATM cells (as discussed in Section 7.2). In addition, they should be simple enough to be implemented in real time.

Let X be an $N_1 \times N_2$ decompressed frame from an MPEG sequence and let Y be the received version at the output of the channel that may have missing data. Each transmitted picture consists of M macroblocks that have $N \times N$ pixels . Let $\mathbf{x}_i$ be the lexicographic ordering of the ith macroblock in X. The vector $\mathbf{x}$ is then defined to be the concatenation of $\mathbf{x}_1, \mathbf{x}_2, \ldots, \mathbf{x}_M$, that is

$$\mathbf{x} = [\mathbf{x}_1^T \ \mathbf{x}_2^T \ \ldots \ \mathbf{x}_M^T]^T.$$

The vector $\mathbf{y}$ is similarly defined for Y. If the jth macroblock is missing due to ATM cell loss then

$$\mathbf{y} = \mathbf{Dx},$$

where $\mathbf{D}$ is an $(N_1 N_2 - N^2) \times N_1 N_2$ matrix that consists of the identity matrix excluding the rows from row jN^2 to row $(j + 1)N^2 - 1$. If n of the M macroblocks are missing due to ATM cell loss then $\mathbf{D}$ will be an $(N_1 N_2 - nN^2) \times N_1 N_2$ matrix. The goal of error concealment is to estimate $\mathbf{x}$ given the received data $\mathbf{y}$.

7.5 DETERMINISTIC SPATIAL APPROACH

Suppose that the kth macroblock, $\mathbf{x}_k$ has been lost during the process of transmission. Let $\hat{X}_{i,j}$ denote the reconstructed value of the sample at the ith row and j^{th} column of $\mathbf{x}_k$, and let $\mathbf{J}_k$ denote the set of indices of the pixels belonging to $\mathbf{x}_k$, that is

$$\mathbf{J}_k = \{(i,j) \mid X_{i,j} \in \mathbf{x}_k\}.$$

We will discuss two methods for reconstructing a lost macroblocks from its neighbors.

7.5.1 Interpolation

In this method, every pixel in $\mathbf{x}_k$ is reconstructed by spatially averaging the values of its four closest intact neighbors as shown in Figure 7.10.

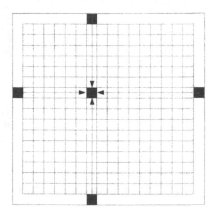

Figure 7.10   Spatial Averaging.   Each lost pixel (shown here in black with arrows pointing to it) is reconstructed from its four closest intact pixels.

$$\hat{X}_{i,j} = \lambda[\mu_1 X_{i,-1} + (1 - \mu_1)X_{i,N}] + (1 - \lambda)[(1 - \mu_2)X_{-1,j} + \mu_2 X_{N,j}],$$

where

$X_{i,-1}$ - is the closest element in the macroblock to the left of x_k,

$X_{i,N}$ - is the closest element in the macroblock to the right of x_k,

$X_{-1,j}$ - is the closest element in the macroblock above x_k,

$X_{N,j}$ - is the closest element in the macroblock below x_k .

The weighting coefficients μ_1 and $1 - \mu_1$ are used to weigh the contributions from the pixels on either side of the lost pixel, and μ_2 and $1 - \mu_2$ are used to weigh the contributions from those above and below the lost pixel. The coefficient μ_1 is a function of the distances between the lost pixel and its closest neighbors lying on the same row, while μ_2 depends on the distances between the lost pixel and its closest neighbors in the same column. The contributions from the macroblocks on either side are weighted by λ, and those from the ones above and below are weighted by $1 - \lambda$. The advantage of this method is that it is simple, fast, and results in good reconstruction.

An alternative approach has been proposed to estimate missing edges in each block from edges in the surrounding blocks [38]. For each direction of an estimated edge, a version of the lost block is reconstructed by performing a number of one dimensional interpolations carried out along several lines parallel to the edge. A final version of the missing block is obtained by merging all previously attained versions together.

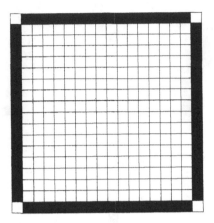

Figure 7.11 The cost function includes the differences between pixels on the boundary of the lost macroblock and their neighbors. These neighboring pixels that belong to the other macroblocks, are shown here in black.

7.5.2 Optimal Iterative Reconstruction

The second proposed method aims at reconstructing the lost macroblocks by minimizing a cost function. This cost function, f, is the sum of the weighted square differences between each lost pixel value and its neighbors [47, 34], which include pixels from surrounding undamaged blocks shown in the dark region in Figure 7.11. Thus,

$$
f(\hat{\mathbf{x}}_{\mathbf{k}}, Y) = \frac{1}{2} \sum_{(i,j)\epsilon \mathbf{J_k}} [\omega_{i,j}^{w}(\hat{X}_{i,j} - \hat{X}_{i,j-1})^2 + \omega_{i,j}^{e}(\hat{X}_{i,j} - \hat{X}_{i,j+1})^2
$$
$$
+ \omega_{i,j}^{n}(\hat{X}_{i,j} - \hat{X}_{i-1,j})^2 + \omega_{i,j}^{s}(\hat{X}_{i,j} - \hat{X}_{i+1,j})^2], \qquad (7.1)
$$

where $\hat{\mathbf{x}}_{\mathbf{k}}$ is the reconstructed version of $\mathbf{x}_{\mathbf{k}}$, and $\omega_{i,j}^{w}$, $\omega_{i,j}^{e}$, $\omega_{i,j}^{n}$ and $\omega_{i,j}^{s}$ are the weighting coefficients for the pixel values left of, right of, above, and below the current pixel, respectively. The explicit dependence of the cost function on Y is used to indicate that the sum in Equation 7.1 contains pixel values from the dark region shown in Figure 7.11. Since the macroblocks in each frame are numbered in lexicographic ordering then Equation 7.1 can be rewritten as

$$
f(\hat{\mathbf{x}}_{\mathbf{k}}, Y) = \frac{1}{2} \sum_{i=\lfloor \mathbf{k}/N \rfloor}^{i=\lfloor \mathbf{k}/N \rfloor + N - 1} \sum_{j=\bmod(\mathbf{k},N)}^{j=\bmod(\mathbf{k},N)+N-1}
$$
$$
[\omega_{i,j}^{w}(\hat{X}_{i,j} - \hat{X}_{i,j-1})^2 + \omega_{i,j}^{e}(\hat{X}_{i,j} - \hat{X}_{i,j+1})^2
$$

$$+ \omega_{i,j}^n (\hat{X}_{i,j} - \hat{X}_{i-1,j})^2 + \omega_{i,j}^s (\hat{X}_{i,j} - \hat{X}_{i+1,j})^2], \qquad (7.2)$$

where

$\lfloor a \rfloor$ - is the greatest integer less than or equal to real number a and

$\mathrm{mod}(\mathbf{k}, N)$ - is the remainder resulting from having divided $\mathbf{k}$ by N.

The same set of weighting coefficients is used to reconstruct each missing macroblock. Without loss of generality, the indices i and j in Equation 7.2 can be made to vary between 1 and N. Hence

$$f(\hat{\mathbf{x}}_\mathbf{k}, Y) = \frac{1}{2} \sum_{i=1}^{i=N} \sum_{j=1}^{j=N} [\omega_{i,j}^w (\hat{X}_{i,j} - \hat{X}_{i,j-1})^2 + \omega_{i,j}^e (\hat{X}_{i,j} - \hat{X}_{i,j+1})^2$$

$$+ \omega_{i,j}^n (\hat{X}_{i,j} - \hat{X}_{i-1,j})^2 + \omega_{i,j}^s (\hat{X}_{i,j} - \hat{X}_{i+1,j})^2], \qquad (7.3)$$

Equation 7.3 can be rewritten as

$$f(\hat{\mathbf{x}}_\mathbf{k}, Y) = \frac{1}{2} \hat{\mathbf{x}}_\mathbf{k}^T \mathbf{Q} \hat{\mathbf{x}}_\mathbf{k} - \hat{\mathbf{x}}_\mathbf{k}^T \mathbf{b} + c,$$

where

$$\mathbf{b} = \sum_{k=1}^{4} [\mathbf{S}_k - \mathbf{Q}_k^T \mathbf{S}_k] \mathbf{b}_k,$$

$$\mathbf{Q} = \sum_{k=1}^{4} [\mathbf{S}_k - \mathbf{S}_k \mathbf{Q}_k - \mathbf{Q}_k^T \mathbf{S}_k + \mathbf{Q}_k^T \mathbf{S}_k \mathbf{Q}_k],$$

$$c = \frac{1}{2} \sum_{k=1}^{4} \mathbf{b}_k^T \mathbf{S}_k \mathbf{b}_k,$$

$$\mathbf{S}_1 = \mathrm{diagonal}(\omega_{1,1}^w, \omega_{1,2}^w, \ldots, \omega_{1,N}^w, \omega_{2,1}^w, \omega_{2,2}^w, \ldots, \omega_{2,N}^w, \ldots, \omega_{N,1}^w, \omega_{N,2}^w, \ldots, \omega_{N,N}^w),$$

$$\mathbf{S}_2 = \mathrm{diagonal}(\omega_{1,1}^e, \omega_{1,2}^e, \ldots, \omega_{1,N}^e, \omega_{2,1}^e, \omega_{2,2}^e, \ldots, \omega_{2,N}^e, \ldots, \omega_{N,1}^e, \omega_{N,2}^e, \ldots, \omega_{N,N}^e),$$

$$\mathbf{S}_3 = \mathrm{diagonal}(\omega_{1,1}^n, \omega_{1,2}^n, \ldots, \omega_{1,N}^n, \omega_{2,1}^n, \omega_{2,2}^n, \ldots, \omega_{2,N}^n, \ldots, \omega_{N,1}^n, \omega_{N,2}^n, \ldots, \omega_{N,N}^n),$$

$$\mathbf{S}_4 = \text{diagonal}(\omega_{1,1}^s, \omega_{1,2}^s, \ldots, \omega_{1,N}^s, \omega_{2,1}^s, \omega_{2,2}^s, \ldots, \omega_{2,N}^s, \ldots, \omega_{N,1}^s, \omega_{N,2}^s, \ldots, \omega_{N,N}^s),$$

$$\mathbf{b}_1 = [X_{1,0}\, 0\, 0 \, \ldots \, 0\, X_{2,0}\, 0\, 0 \, \ldots \, 0 \, \ldots \, X_{N,0}\, 0\, 0 \, \ldots \, 0]^T,$$

$$\mathbf{b}_2 = [0\, 0\, 0 \, \ldots \, X_{1,N+1}\, 0\, 0\, 0 \, \ldots \, X_{2,N+1} \, \ldots \, 0\, 0\, 0 \, \ldots \, X_{N,N+1}]^T,$$

$$\mathbf{b}_3 = [X_{0,1}\, X_{0,2}\, X_{0,3} \, \ldots \, X_{0,N}\, 0\, 0\, 0 \, \ldots \, 0 \, \ldots \, 0\, 0\, 0 \, \ldots \, 0]^T,$$

and

$$\mathbf{b}_4 = [0\, 0\, 0 \, \ldots \, 0\, 0\, 0\, 0 \, \ldots \, 0 \, \ldots \, X_{N+1,1}\, X_{N+1,2}\, X_{N+1,3} \, \ldots \, X_{N+1,N}]^T,$$

where diagonal($\cdot$) indicates a diagonal matrix with the arguments comprising the diagonal elements of the matrix. The matrices $\mathbf{Q}_k$ are upper and lower diagonal matrices with zeros along the diagonals that satisfy

$$[\mathbf{Q}_1]_{i,j} = \begin{cases} 1 & i = j + 1 \text{ and } \text{mod}\,(i, N) \neq 1 \\ 0 & \text{otherwise,} \end{cases}$$

$$[\mathbf{Q}_3]_{i,j} = \begin{cases} 1 & i = j + N \\ 0 & \text{otherwise,} \end{cases}$$

$$\mathbf{Q}_2^T = \mathbf{Q}_1,$$

and

$$\mathbf{Q}_4^T = \mathbf{Q}_3.$$

When all weighting coefficients are greater than zero then, $\mathbf{Q}$ is positive definite, which is necessary for attaining the optimal solution. In this case the optimal solution, in the mean square sense, is

$$\hat{\mathbf{x}}_k^{opt} = \mathbf{Q}^{-1}\mathbf{b}.$$

The above equation can be iteratively solved. For instance, the following equations depict the lth iteration for $\mathbf{x}$ in the event that the steepest descent algorithm is used to find $\mathbf{x}$,

$$\mathbf{x}_l = \mathbf{x}_{l-1} - \frac{\mathbf{g}_{l-1}^T \mathbf{g}_{l-1}}{\mathbf{g}_{l-1}^T \mathbf{Q} \mathbf{g}_{l-1}} \mathbf{g}_{l-1},$$

where

$$\mathbf{g}_l = \mathbf{Q}\mathbf{x}_l - \mathbf{b},$$

is the lth iteration of the gradient vector of f.

7.6 STATISTICAL SPATIAL APPROACH: MAP ESTIMATION

Recently, techniques for edge reconstruction [48, 49, 50, 51, 52] have gained popularity. The underlying idea is that the original image is modeled as a Markov Random Field (MRF) [53, 54, 48]. Edges are then reconstructed by maximum *a posteriori* (MAP) techniques. This is the approach adopted here. Each original frame X and its received version Y are modeled as discrete parameter random fields where each pixel is a continuous random variable. Assuming a prior distribution for $\mathbf{x}$, a maximum *a posteriori* (MAP) estimate is obtained given the received data $\mathbf{y}$. Denoting the estimate of $\mathbf{x}$ by $\hat{\mathbf{x}}$,

$$\hat{\mathbf{x}} = \arg \max_{\mathbf{x}|\mathbf{y}=\mathbf{Dx}} L(\mathbf{x} \mid \mathbf{y}),$$

where $L(\mathbf{x} \mid \mathbf{y})$ is the log-likelihood function. In other words,

$$L(\mathbf{x} \mid \mathbf{y}) = \ln f(\mathbf{x} \mid \mathbf{y}),$$

where $f(\mathbf{x} \mid \mathbf{y})$ is the conditional probability density function of $\mathbf{x}$ given $\mathbf{y}$. It can be shown [41] that

$$\hat{\mathbf{x}} = \arg \min_{\mathbf{x}|\mathbf{y}=\mathbf{Dx}} [-\ln f(\mathbf{x})].$$

Since X is modeled as a Markov Random Field (MRF), the probability density function of $\mathbf{x}$ is given by [54, 48, 55]

$$f(\mathbf{x}) = \frac{1}{Z} \exp\left(-\sum_{c \epsilon C} V_c(\mathbf{x})\right),$$

where Z is a normalizing constant known as the *partition function*, $V_c(\cdot)$ a function of a local group of points c known as cliques, and C the set of all cliques [48].

The MAP estimate of $\mathbf{x}$ is then given by

$$\hat{\mathbf{x}} = \arg \min_{\mathbf{x}|\mathbf{y}=\mathbf{Dx}} \left[\sum_{c \epsilon C} V_c(\mathbf{x})\right].$$

7.6.1 Implementation

The proper choice of V_c is crucial to the reconstruction of the macroblocks. In this case the potential functions are

$$\sum_{c \epsilon C} V_c(\mathbf{x}) = \sum_{i=0}^{N_1-1} \sum_{j=0}^{N_2-1} \sum_{m=0}^{3} b_{i,j}^{(m)} \rho(\frac{D_m(X_{i,j})}{\sigma}),$$

where

$$
\begin{aligned}
D_0(X_{i,j}) &= X_{i,j-1} - X_{i,j}, \\
D_1(X_{i,j}) &= X_{i-1,j+1} - X_{i,j}, \\
D_2(X_{i,j}) &= X_{i-1,j} - X_{i,j}, \\
D_3(X_{i,j}) &= X_{i-1,j-1} - X_{i,j},
\end{aligned}
$$

are used to approximate the first order derivatives at the $i^{th}j^{th}$ pixel. $\rho(\cdot)$ is a cost function, σ a scaling factor, $b_{i,j}^{(m)}$ weighting coefficients, and the set of cliques [48] is

$$
C = \{ \ \{(i, j-1), (i, j)\}, \{(i-1, j+1), (i, j)\},
$$
$$
\{(i-1, j), (i, j)\}, \{(i-1, j-1), (i, j)\} \ \}.
$$

Several cost functions have been proposed [49, 51]. A convex $\rho(\cdot)$ results in the minimization of a convex functional. The cost function used here is the one introduced by Huber [56]. It is defined to be

$$
\rho_\gamma(x) = \begin{cases} x^2 & |x| \leq \gamma \\ \gamma^2 + 2\gamma(|x| - \gamma) & |x| > \gamma. \end{cases}
$$

Hence,

$$
\sum_{c \in C} V_c(\mathbf{x}) = \sum_{i=0}^{N_1-1} \sum_{j=0}^{N_2-1} \sum_{m=0}^{3} b_{i,j}^m \rho_\gamma \left(\frac{D_m(X_{i,j})}{\sigma} \right).
$$

Let

$$
h_\gamma(\mathbf{x}) = \sum_{i=0}^{N_1-1} \sum_{j=0}^{N_2-1} \sum_{m=0}^{3} b_{i,j}^m \rho_\gamma \left(\frac{D_m(X_{i,j})}{\sigma} \right),
$$

then

$$
\hat{\mathbf{x}} = \arg \min_{\mathbf{x}|\mathbf{y}=\mathbf{D}\mathbf{x}} h_\gamma(\mathbf{x}).
$$

The solution to the equation above can be obtained by means of the iterative conditional modes (ICM) algorithm [57]. In particular if the ith element of $\mathbf{x}$ corresponds to a lost pixel value and $\mathbf{x}_{\partial i}$ denotes the neighborhood of x_i then

$$
\hat{x}_i = \arg \max_{x_i} f(x_i \mid \mathbf{x}_{\partial i}). \tag{7.4}
$$

It can be shown that the the MAP estimate of pixel (i, j) given its neighbors is

$$
\hat{X}_{i,j} = \arg \min_{X_{i,j}} \sum_{l=i}^{i+1} \sum_{k=j}^{j+1} \sum_{m=0}^{3} b_{l,k}^m \rho_\gamma \left(\frac{D_m(X_{l,k})}{\sigma} \right). \tag{7.5}
$$

Let $\mathbf{X}_{\partial i}$ denote the neighborhood of the ith macroblock $\mathbf{x_i}$. The MAP estimate of $\mathbf{x_i}$ satisfies

$$\hat{\mathbf{x}}_i = \arg\max_{\mathbf{x}} f(\mathbf{x_i} \mid \mathbf{X}_{\partial i}).$$

If we let $\mathbf{J_i}$ denote the set of indices of the pixels belonging to $\mathbf{x_i}$, then it can be similarly shown that

$$\hat{\mathbf{x}}_i = \arg\min_{\mathbf{x_i}} \sum_{(i,j)\epsilon J_i} \sum_{l=i}^{i+1} \sum_{k=j}^{j+1} \sum_{m=0}^{3} b_{l,k}^m \rho_\gamma\left(\frac{D_m(X_{l,k})}{\sigma}\right). \tag{7.6}$$

The solution to Equation (7.6) can be obtained iteratively.

7.7 A SUBOPTIMAL APPROACH

The choice of γ and σ is crucial to the reconstruction of edges. The smaller the product $\gamma\sigma$, the less edges are penalized. Here σ is chosen to be of unit value. In addition the weighting coefficients $b_{i,j}^m$ are also chosen to be of unit value. Thus,

$$h_\gamma(\mathbf{x}) = \sum_{i=0}^{N_1-1} \sum_{j=0}^{N_2-1} \sum_{m=0}^{3} \rho_\gamma(D_m(X_{i,j})).$$

Now,

$$\begin{aligned}
\frac{\partial}{\partial X_{i,j}} h_\gamma(\mathbf{x}) &= \rho_\gamma'(D_m(X_{i,j+1})) + \rho_\gamma'(D_m(X_{i+1,j-1})) \\
&\quad + \rho_\gamma'(D_m(X_{i+1,j})) + \rho_\gamma'(D_m(X_{i+1,j+1})) \\
&\quad - \sum_{m=0}^{3} \rho_\gamma'(D_m(X_{i,j})),
\end{aligned}$$

where

$$\rho_\gamma'(x) = \begin{cases} 2x & |x| \leq \gamma \\ 2\gamma & x > \gamma \\ -2\gamma & x < -\gamma. \end{cases} \tag{7.7}$$

Since $h_\gamma(\mathbf{x})$ is continuous, convex, and has continuous first partial derivatives, then by successively iterating with respect to each pixel, a global minimum is attained. However, there is more than one global minimum.

Each pixel in the interior, has 8 neighbors. Let $z_1, z_2, z_3, z_4, z_5, z_6, z_7, z_8$ be the 8 neighbors arranged in ascending order, and let

$$\begin{aligned}
U = \ & \{(k,l) \mid k = i-1 \text{ and } l = j-1, j, j+1 \\
& \text{or } k = i \text{ and } l = j-1\},
\end{aligned} \tag{7.8}$$

$$L = \{(k, l) \mid k = i + 1 \text{ and } l = j - 1, j, j + 1$$
$$\text{or } k = i \text{ and } l = j + 1\}, \tag{7.9}$$

and

$$\Delta_k(x) = \begin{cases} z_k - X_{i,j} & z_k \epsilon U \\ X_{i,j} - z_k & z_k \epsilon L. \end{cases}$$

Since, we are iterating for $X_{i,j}$, we need to solve

$$\frac{\partial}{\partial X_{i,j}} h_\gamma(\mathbf{x}) = 0, \tag{7.10}$$

When solving Equation (7.10), three cases need to be considered.

Case1: $|\Delta_k| \leq \gamma \ \forall k$

This occurs when $z_8 - \gamma \leq z_1 + \gamma$, and the optimum value of $X_{i,j}$ satisfies the constraint

$$|\Delta_k| \leq \gamma \ \forall k.$$

In this case using Equations (7.10) and (7.7)

$$\hat{X}_{i,j} = \frac{1}{8} \sum_{k=1}^{8} z_k.$$

Case2: $|\Delta_k| > \gamma \ \forall k$

This occurs when the optimum value of $X_{i,j}$ satisfies

$$X_{i,j} > z_k + \gamma \text{ or } X_{i,j} < z_k - \gamma \ \forall k$$

in addition to Equation (7.10). In this case Equation (7.10) yields

$$\sum_{(k,l)\epsilon L} \beta_{k,l} + \sum_{(k,l)\epsilon U} \beta_{k,l} = 0, \tag{7.11}$$

where

$$\beta_{k,l} = \begin{cases} 1 & X_{i,j} > X_{k,l} + \gamma \\ -1 & X_{i,j} < X_{k,l} - \gamma. \end{cases}$$

Equation (7.11) is satisfied only if the following holds

$$z_1 \leq z_2 \leq z_3 \leq z_4 < z_4 + \gamma < \hat{X}_{i,j} < z_5 - \gamma < z_5 \leq z_6 \leq z_7 \leq z_8.$$

One possible choice for $\hat{X}_{i,j}$ is

$$\hat{X}_{i,j} = \text{median}\{z_4, z_5\}.$$

Case3:

Let

$$I = \{k| \; |\Delta_k| \leq \gamma\},$$
$$I^c = \{k| \; |\Delta_k| > \gamma\},$$

and

$$\alpha_k = \begin{cases} 1 & k\epsilon I \\ 0 & k\epsilon I^c. \end{cases}$$

From Equations (7.10) and (7.7)

$$\hat{X}_{i,j} = \frac{\sum_{k\epsilon I} z_k - \gamma \sum_{k\epsilon I^c} \beta_k}{\sum_{k\epsilon I} \alpha_k}.$$

By choosing γ arbitrarily small and positive, the optimum value for $X_{i,j}$ will satisfy the second case unless there are at least two of the neighboring pixels that are equal in value. Under such conditions, the value of the common pixel value is used. Although this is a suboptimal strategy, the resulting reconstruction technique is faster than when line search techniques such as Steepest Descent or Newton-Raphson [58] are utilized to attain the optimum values.

7.8 TEMPORAL RESTORATION: MOTION VECTOR ESTIMATION

Most frames in an MPEG sequence are predicted frames that have motion vectors associated with their macroblocks by which they are reconstructed at the decoder. A more expedient way of reconstructing a lost macroblock would be to estimate its associated missing motion vector.

7.8.1 Deterministic

One approach is to average the motion vectors of surrounding macroblocks [23, 25] and use the average vector to retrieve a version of the lost macroblock. This retrieved macroblock is then averaged with another version obtained via spatial interpolation.

7.8.2 Statistical

An alternative approach is based on the MRF model. Let v_i be the motion vector associated with the ith macroblock in the current frame. In lossless transmission, $x_i^{(0)}$, the ith macroblock in the current frame, is reconstructed by the decoder as

$$x_i^{(0)} = x_{i-v_i}^{(-1)} + n_i.$$

Here $x_{i-v_i}^{(-1)}$ is the macroblock in the previous frame that closely matches $x_i^{(0)}$, and n_i is the decoded error arising from having replaced $x_i^{(0)}$ by $x_{i-v_i}^{(-1)}$.

In a lossy transmission it is not possible to recover $\mathbf{n}_i$. The aim is then to obtain an estimate for $\mathbf{v}_i$ that will point to $\mathbf{x}_{i-\mathbf{v}_i}^{(-1)}$. In this case the estimate $\hat{\mathbf{v}}_i$ is posed to satisfy

$$\hat{\mathbf{v}}_i = \arg\max_{\mathbf{v}_i} f(\mathbf{x}_{i-\mathbf{v}_i}^{(-1)} \mid \mathbf{X}_{\partial i}^{(0)}),$$

where $\mathbf{X}_{\partial i}^{(0)}$ is the intact neighborhood of $\mathbf{x}_i{}^{(0)}$. Alternatively,

$$\hat{\mathbf{v}}_i = \arg\min_{\mathbf{v}_i} \sum_{(r,s)\mid X_{(r,s)}\epsilon\mathbf{x}_{i-\mathbf{v}_i}^{(-1)}} \sum_{l=r}^{r+1}\sum_{k=s}^{s+1}\sum_{m=0}^{3} b_{l,k}^m \rho_\gamma\left(\frac{D_m(X_{l,k})}{\sigma}\right). \tag{7.12}$$

The optimal estimate is obtained by searching for the vector that minimizes Equation (7.12). Rather than searching the entire image for $\hat{\mathbf{v}}_i$, the motion vectors associated with the neighboring macroblocks can be used to bound the search space.

7.9   SUBOPTIMAL MOTION VECTOR ESTIMATION

A suboptimal approach to estimating $\mathbf{v}_i$, the motion vector associated with the ith macroblock, is based on evaluating Equation (7.12) for the pixels on the boundary, that is

$$\hat{\mathbf{v}}_i = \arg\min_{\mathbf{v}_i} \sum_{(r,s)\mid X_{(r,s)}\epsilon\mathbf{B}_{i-\mathbf{v}_i}^{(-1)}} \sum_{l=r}^{r+1}\sum_{k=s}^{s+1}\sum_{m=0}^{3} b_{l,k}^m \rho_\gamma\left(\frac{D_m(X_{l,k})}{\sigma}\right). \tag{7.13}$$

where $\mathbf{B}_{i-\mathbf{v}_i}^{(-1)}$ consists of the pixels lying on the boundary of the macroblock $\mathbf{x}_{i-\mathbf{v}_i}^{(-1)}$, as illustrated in Figure 7.12. This is equivalent to finding the MAP estimate of the boundary pixels of the lost macroblock and not the MAP estimate of the entire lost macroblock. In view of the discussion of Section (7.7), Equation (7.13) can be rewritten as

$$\hat{\mathbf{v}}_i = \arg\min_{\mathbf{v}_i} \sum_{(r,s)\mid X_{(r,s)}\epsilon\mathbf{B}_{i-\mathbf{v}_i}^{(-1)}} \sum_{l=r}^{r+1}\sum_{k=s}^{s+1}\sum_{m=0}^{3} |D_m(X_{l,k})|. \tag{7.14}$$

The estimate of the missing motion vector is then obtained by searching for the vector that minimizes Equation (7.14). If the search process yields more than one estimate, the motion vector having the shortest length is then chosen.

7.10   RESULTS

To test the algorithms, the *jane* and *salesman* sequences were packed into ATM cells as described in Section 7.2 and two percent of the cells dropped.

Figure 7.13a is a decoded frame from the *salesman* sequence. Shown in Figure 7.13b is the same frame with some of its macroblocks missing. In Figure 7.13c the

☐   Prospective macroblock
▨   Neighbors of lost macroblock
■ $B_{i-v_i}^{(-1)}$

Figure 7.12   Boundary pixels of prospective macroblock

deterministic spatial interpolation algorithm (Section 7.5.2) was used to reconstruct the missing data.

Figure 7.13   Deterministic spatial interpolation of lost data. The figure in (a) is a decoded frame from the *salesman* sequence. Depicted in (b) is a version with randomly missing macroblocks.

Figure 7.14b is a frame missing macroblocks, that were reconstructed in Figure 7.14c. To reconstruct each missing macroblock, the MAP estimate of each pixel, based on Equation (7.4), was obtained. After estimating all pixel values of a macroblock, the entire procedure was repeated a number of times. A total of two iterations was sufficient to reconstruct the missing macroblock. The original decoded frame is provided in Figure 7.14a.

(a)　　　　　　　　　　　(b)

(c)

Figure 7.14　Spatial reconstruction using MAP estimation. The figure in (a) is a decoded frame from the *salesman* sequence. It is shown in (b) to be missing some of its macroblocks, that are replaced in (c) by their MAP estimates.

Figure 7.15a is a frame from the *jane* sequence. Figure 7.15c shows the reconstructed version using our spatial approach based on median filtering, and Figure 7.15d shows the same frame reconstructed by means of the iterative technique proposed in [41]. In both cases, the missing macroblocks were reconstructed from spatial data. The advantage of the current approach is that a MAP estimate is obtained much faster, making this approach attractive for real-time implementation on set-top decoder boxes.

Figure 7.16a is a decoded frame from the *salesman* sequence. Due to cell loss the major portions the frame were lost as shown in Figure 7.16b. Reconstruction is performed by searching for the motion vector that minimizes Equation (7.12). As seen the reconstructed version in Figure 7.16c closely matches the original in Figure 7.16a.

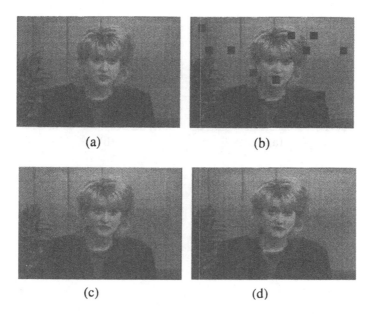

(a)   (b)

(c)   (d)

Figure 7.15   Spatial reconstruction based on median filtering. The figure in (a) is a decoded frame from the *jane* sequence. It is shown in (b) to be missing some of its macroblocks, that were reconstructed using median filtering. The same macroblocks are restored in (d) by using iterating for the MAP estimates using line search techniques.

Figure 7.17a is a frame from the *salesman* sequence that has been damaged in Figure 7.17b and restored in Figure 7.17c. Reconstruction was carried out by temporal replacement from the previous frame. In this case the motion vector was estimated by averaging the vectors belonging to neighboring macroblocks. Due to the fact that the neighbors contained no motion, the vector pointed to the same macroblock in the previous frame.

A hybrid of temporal and spatial interpolation can also be used. Figure 7.18b, is a predicted frame from the the *salesman* sequence that has also been corrupted due to cell loss and Figure 7.18c is the restored version using motion vector estimation based on Equation (7.12). In Figure 7.18d the spatial interpolation algorithm was applied to Figure 7.18c for further restoration. The search space for the motion vectors did not exceed an area of 16x16 pixels and hence an exhaustive search for the motion vector was implemented. In the case of sequences where a substantial amount of motion exits an exhaustive search may be too costly and thus other searches such as the logarithmic search may be implemented but at a cost of lower fidelity.

Figure 7.19a is a frame from the *jane* sequence, and Figure 7.19b is the same frame with missing macroblocks due to cell loss. In Figure 7.19c the lost macroblocks

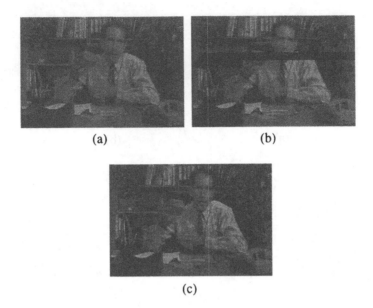

Figure 7.16 Temporal reconstruction based on the MRF model of the frames. The figure in (a) is a decoded frame from the *salesman* sequence, and that in (b) is a damaged version due to ATM cell loss. The reconstructed version, shown in, (c), was obtained by first estimating the missing motion vectors and then using the estimates to fill in the lost pixel data.

were reconstructed by obtaining the suboptimal estimates of their motion vectors as described in Section 7.9. This is compared to Figure 7.19d which illustrates the reconstruction of the same frame based on Equation (7.12).

7.11 SUMMARY

We have presented both spatial and temporal error concealment techniques for coded video. The spatial techniques reconstruct a missing pixel value in a video frame by interpolating it from neighboring macroblocks. This can be done by either minimizing intersample variations within each missing macroblock and across its boundaries, or by modeling each frame as a Markov random field and obtaining a MAP estimate of the missing pixel values. We also described a suboptimal approach to finding the MAP estimate based on median filtering. Temporal restoration techniques based on finding missing motion vectors were also presented. Missing motion vectors can be estimated either by maximizing the probability density function of the missing macroblock given its neighbors, or by simply averaging the motion vectors of the neighboring macroblocks.

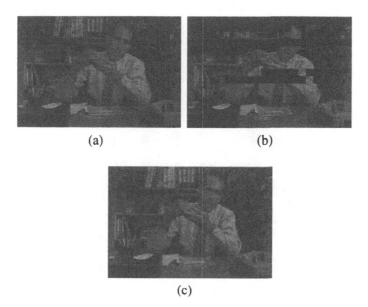

(a) (b)

(c)

Figure 7.17 Reconstruction of lost data based on temporal replacement from the previous frame. The figure in (a) is a decoded frame from the *salesman* sequence, and that in (b) is a damaged version due to ATM cell loss. The reconstructed version, shown in, (c), was obtained by replacing the lost macroblocks with macroblocks from the preceding frame that have the same address.

References

[1] J. Watkinson, *The art of digital video*. Oxford, England: Focal Press, second ed., 1994.

[2] C. P. Sandbank, *Digital Television*. John Wiley and Sons, 1992.

[3] A. M. Tekalp, *Digital Video Processing*. Prentice Hall, 1995.

[4] A. A. Rodriguez, C. E. Fogg, and E. J. Delp, "Video compression for multimedia applications," in *Image Technology: Advances in Image Processing, Multimedia and Machine Vision*, pp. 613–679, Springer-Verlag, 1996.

[5] MPEG–1 Standard, *ISO/IEC 11172, Information Technology–coding of moving pictures and associated audio for digital storage media at up to about 1. 5 Mbits/s*. ISO, 1993.

[6] MPEG–2 Standard, *ISO/IEC 13818, Generic Coding of Moving Pictures and Associated Audio Information*. ISO, 1995.

[7] D. LeGall, "MPEG: A video compression standard for multimedia applications," *Communications of the ACM*, vol. 34, no. 4, pp. 47–58, April 1991.

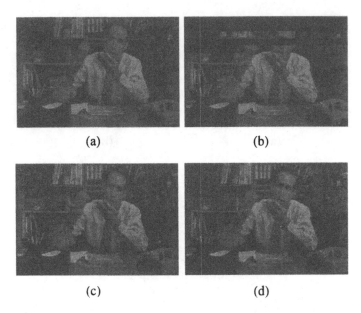

(a)   (b)

(c)   (d)

Figure 7.18   Temporal reconstruction based on the MRF model followed by smoothing of reconstructed data using MAP estimation. Shown in (a) is a decoded frame from the *salesman* sequence. The frame was damaged due to ATM cell loss, shown in (b), and reconstructed in (c) and (d). In (c) the frame was restored by estimating the missing motion vectors and then using them to reconstruct the lost data. Shown in (d) is the outcome of having applied the spatial technique for reconstruction to the image in (c).

[8] V. Bhaskaran and K. Konstantinides, *Image and Video Compression Standards: Algorithms and Architectures.* Kluwer Academic Publisher, 1995.

[9] J. L. Black, W. B. Pennebaker, C. E. Fogg, and D. J. LeGall, *MPEG Video Compression Standard.* Digital Multimedia Standards Series, Chapman and Hall, 1996.

[10] K. R. Rao and J. J. Hwang, *Techniques and Standards for Image, Video and Audio Coding.* Prentice Hall, 1996.

[11] Requirements, Audio, DMIF, SNHC, Systems, Video, "Overview of the MPEG–4 standard." Stockholm meeting, document ISO/IEC JTC1/SC29/WG11 N1730, July 1997, July 1997.

[12] MPEG Requirements Group, "MPEG–7: Context and objectives (version 4)." Stockholm meeting, document ISO/IEC JTC1/SC29/WG11 N1733, July 1997, July 1997.

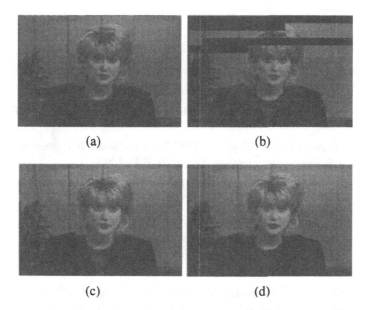

(a)   (b)

(c)   (d)

Figure 7.19   Reconstruction of lost data based on suboptimal temporal replacement from the previous frame. The figure in (a) is a decoded frame from the *jane* sequence and that in (b) is a damaged version due to ATM cell loss.   In (c) the frame was restored by finding suboptimal estimates of the missing motion vectors (c). In (d), error concealment was achieved by finding optimal estimates for the missing motion vectors.

[13] A. K. Jain, *Fundamentals of digital image processing.* Prentice Hall Information and System Sciences Series, Prentice Hall, 1989.

[14] MPEG–1 Video Coding Standard, *ISO/IEC 11172–2, Information Technology– coding of moving pictures and associated audio for digital storage media at up to about 1. 5 Mbits/s–Part2: Video.* ISO, 1993.

[15] J. R. Jain and A. K. Jain, "Displacement measurement and its application in interframe image coding," *IEEE Transactions on Communications*, vol. COM-29, no. 12, pp. 1799–1808, December 1981.

[16] MPEG–1 Systems Standard, *ISO/IEC 11172–1, Information Technology–coding of moving pictures and associated audio for digital storage media at up to about 1. 5 Mbits/s–Part1: Systems.* ISO, 1993.

[17] MPEG–2 Systems Standard, *ISO/IEC 13818–1, Generic Coding of Moving Pictures and Associated Audio Information–Part1: Systems.* ISO, 1995.

[18] B. G. Haskell, A. Puri, and A. N. Netravali, *Digital Video: An Introduction to MPEG-2.* Digital Multimedia Standards Series, Chapman and Hall, 1997.

230

[19] G. C. Kessler and P. Southwick, *ISDN Concepts, Facilities and Services*. McGraw-Hill Series on Computer Communications, McGraw-Hill, third ed., 1997.

[20] U. Black, *ATM:Foundation for broadband networks*. Prentice Hall Series in Advanced Communications Technology, Prentice Hall, 1995.

[21] V. Kumar, *Broadband communications*. McGraw-Hill Series on Computer Communications, McGraw-Hill, 1995.

[22] L. G. Cuthbert and J.-C. Sapanel, *ATM:The broadband telecommunications solution*. IEE Telecommunications Series 29, IEE, 1993.

[23] M. Wada, "Selective recovery of video packet loss using error concealment," *IEEE Journal on Selected Areas in Communication*, vol. 7, no. 5, pp. 807–814, June 1989.

[24] F. Kishino, K. Matanabe, Y. Hayashi, and H. Yasuda, "Variable bit rate coding of video signals for ATM networks," *IEEE Journal on Selected Areas in Communications*, vol. 7, no. 5, pp. 801–806, June 1989.

[25] M. Ghanbari and V. Seferidis, "Cell-loss concealment in ATM networks," *IEEE Transactions on Circuits and Systems for Video Technology*, vol. 3, no. 3, pp. 238–247, June 1993.

[26] M. Ghanbari and C. Hughes, "Packing coded video signals into ATM cells," *IEEE/ACM Transactions on Networking*, vol. 1, no. 5, pp. 505–508, October 1993.

[27] P. Pancha and M. El Zarki, "MPEG coding for variable bit rate video transmission," *IEEE Communications Magazine*, vol. 32, no. 5, pp. 54–66, May 1994.

[28] W. Luo and M. El Zarki, "Analysis of error concealment schemes for MPEG-2 video transmission over ATM based networks," *Proceedings of the SPIE Conference on Visual Communications and Image Processing*, vol. 2501/3, May 1995, Taipei, Taiwan, pp. 1358–1368.

[29] L. H. Kieu and K. N. Ngan, "Cell loss concealment techniques for layered video codec in an ATM network," *Computers and Graphics: Image Communication*, vol. 18, no. 1, pp. 11–19, January-February 1994.

[30] D. Raychaudhuri, H. Sun, and R. S. Girons, "ATM transport and cell-loss concealment techniques for MPEG video," *Proceedings of the International Conference on Acoustics, Speech and Signal Processing*, November 1993, Minneapolis, Minnesota, pp. 117–120.

[31] A. S. Tom, C. L. Yeh, and F. Chu, "Packet video for cell loss protection using deinterleaving and scrambling," *Proceedings of the International Conference on Acoustics, Speech and Signal Processing*, May 1991, Toronto, Canada, pp. 2857–2860.

[32] Q. Zhu, Y. Wang, and L. Shaw, "Coding and cell loss recovery in DCT based packet video," *IEEE Transactions on Circuits and Systems for Video Technology*, vol. 3, no. 3, pp. 248–258, June 1993.

[33] J.Y.Park, M.H.Lee, and K.J.Lee, "A simple concealment for ATM bursty cell loss," *IEEE Transactions on Consumer Electronics*, vol. 39, no. 3, pp. 704–710, August 1993.

[34] Y. Wang, Q. Zhu, and L. Shaw, "Maximally smooth image recovery in transform coding," *IEEE Transactions on Communications*, vol. 41, no. 10, pp. 1544–1551, October 1993.

[35] Y. Wang and Q. Zhu, "Signal loss recovery in DCT-based image and video codecs," *Proceedings of the SPIE Conference on Visual Communications and Image Processing*, vol. 1605, November 1991, Boston, Massachusetts, pp. 667–678.

[36] L. T. Chia, D. J. Parish, and J. W. R. Griffiths, "On the treatment of video cell loss in in the transmission of motion-JPEG and JPEG images," *Computers and Graphics: Image Communication*, vol. 18, no. 1, pp. 11–19, January-February 1994.

[37] H. Sun and J. Zdepski, "Adaptive error concealment algorithm for MPEG compressed video," *Proceedings of the SPIE Conference on Visual Communications and Image Processing*, vol. 1818, November 1992, Boston, Massachusetts, pp. 814–824.

[38] W. Kwok and H. Sun, "Multidirectional interpolation for spatial error concealment," *IEEE Transactions on Consumer Electronics*, vol. 3, no. 39, pp. 455–460, August 1993.

[39] H. Sun and W. Kwok, "Concealment of damaged block transform coded images using projections onto convex sets," *IEEE Transactions on Image Processing*, vol. 4, no. 4, pp. 470–477, April 1995.

[40] P. Salama, N. Shroff, E. J. Coyle, and E. J. Delp, "Error concealment techniques for encoded video streams," *Proceedings of the International Conference on Image Processing*, October 23–26 1995, Washington, DC, pp. 9–12.

[41] P. Salama, N. Shroff, and E. J. Delp, "A bayesian approach to error concealment in encoded video streams," *Proceedings of the International Conference on Image Processing*, September 16–19 1996, Lausanne, Switzerland, pp. 49–52.

[42] S. Aign and K. Fazel, "Error detection and concealment measures in MPEG–2 video decoder," *Proceedings of the International Workshop on HDTV*, October 1994, Torino, Italy.

[43] S. Aign and K. Fazel, "Temporal and spatial error concealment techniques for hierarchical MPEG–2 video codec," *Proceedings of the International Workshop on Communications*, June 18–20 1995, Seattle, Washington, pp. 1778–1783.

[44] S. Aign, "Error concealment enhancement by using the reliability outputs of a SOVA in MPEG-2 video decoder," *Proceedings of the URSI International Symposium on Signal, Systems, and Electronics*, October 25–27 1995, San Francisco, California, pp. 59–62.

[45] J. F. Shen and H. M. Hang, "Compressed image concealmnet and postprocessing for digital video recording," *Proceedings of the IEEE Asia-Pacific Conference on Circuits and Systems*, December 5–8 1994, Taipei, Taiwan, pp. 636–641.

[46] D. C. Youla and H. Webb, "Image restoration by the method of convex projections: Part 1 - theory," *IEEE Transactions on Medical Imaging*, vol. MI-1, no. 2, pp. 81–94, October 1982.

[47] W. Grimson, "An implementation of a computational theory of visual surface interpolation," *Computer Vision, Graphics, Image Processing*, vol. 22, no. 1, pp. 39–69, April 1983.

[48] S. Geman and D. Geman, "Stochastic relaxation, gibbs distributions, and the bayesian restoration of images," *IEEE Transactions on Pattern Analysis and Machine Intelligence*, vol. PAMI-6, no. 6, pp. 721–741, November 1984.

[49] R. L. Stevenson, B. E. Schmitz, and E. J. Delp, "Discontinuity preserving regularization of inverse visual problems," *IEEE Transactions on Systems Man and Cybernetics*, vol. 24, no. 3, pp. 455–469, March 1994.

[50] D. Geman and G. Reynolds, "Constrained restoration and the recovery of discontinuities," *IEEE Transactions on Pattern Analysis and Machine Intelligence*, vol. 14, no. 3, pp. 367–382, March 1992.

[51] C. Bouman and K. Sauer, "A generalized gaussian image model for edge-preserving MAP estimation," *IEEE Transactions on Image Processing*, vol. 2, no. 3, pp. 296–310, July 1993.

[52] J. Marroquin, S. Mitter, and T. Poggio, "Probablistic solution of ill-posed problems in computational vision," *Journal of the American Statistical Association*, vol. 82, no. 397, pp. 76–89, March 1987.

[53] R. Kinderman and J. L. Snell, *Markov Random Fields and their Applications*, vol. 1 of *Contemporary Mathematics*. American Mathematical Society, 1980.

[54] J. Besag, "Spatial interaction and the statistical analysis of lattice systems," *Journal of the Royal Statistical Society*, series B, vol. 36, pp. 192–326, 1974.

[55] R. Schultz and R. L. Stevenson, "A bayesian approach to image expansion for improved definition," *IEEE Transactions on Image Processing*, vol. 3, no. 3, pp. 233–241, May 1994.

[56] P. J. Huber, *Robust Statistics*. John Wiley & Sons, 1981.

[57] J. Besag, "On the statistical analysis of dirty pictures," *Journal of the Royal Statistical Society*, series B, vol. 48, no. 3, pp. 259–302, 1986.

[58] D. G. Luenberger, *Linear and Nonlinear Programming*. Addison Wesley, second ed., 1989.

8 ERROR CONCEALMENT FOR MPEG-2 VIDEO

Susanna Aign

German Aerospace Center (DLR)
Institute for Communications Technology
D-82230 Wessling, Germany
Susanna.Aign@dlr.de

Abstract: In this chapter the problem of error concealment for MPEG-2 compressed video bitstreams applied to the digital TV-broadcasting is examined and a number of different error concealment approaches are considered. More specifically, temporal, spatial, and combinations of these approaches are investigated. In addition, a method for faster re-synchronization is examined. Simulation results are shown that compare these approaches both objectively and subjectively for different bit error rates. Furthermore, error concealment techniques and experimental comparisons are shown for MPEG-2 scalable bitstreams. For these simulations it is assumed that the base layer is error-free and the enhancement layer is error-prone.

8.1 INTRODUCTION

The MPEG-2 video source coding algorithm will be considered for digital broadcasting of TV-signals over various transmission media such as satellite, cable, and terrestrial channels. Due to variable length coding (VLC) the video-signal is very sensitive to channel disturbances, therefore, a bit error rate (BER) of less than 10^{-8} has to be guaranteed for acceptable quality of transmission. Because of this a powerful error protection schemes are used in the European TV-Broadcasting Standards. However, in spite of this in the case of bad reception conditions, e.g. deep fades or impulsive noise, residual errors may still occur even in the best protected video signal. These residual errors in the video signal result in horizontal stripes of lost macroblocks or in

235

the loss of entire pictures, thus good quality service cannot be guaranteed. Therefore, postprocessing based error concealment techniques at the receiver are necessary to ameliorate these errors. These techniques can be applied at the receiver without changing the encoder.

The aim of this chapter is to provide an overview of different error concealment techniques for handling residual channel errors at the receiver which take into account the properties of the transmission medium. Error concealment techniques such as temporal or spatial error concealment as well as techniques for scalable bitstreams are analyzed and compared to each other. Additional techniques are also investigated which capitalize on the correctly received data. The results presented using both subjective as well as objective measurements in terms of peak-signal-to-noise-ratio (PSNR)-values.

In order to understand the problem and the mechanism of error concealment, which can the entire transmission system including source coding, multiplexing, and channel coding will be first briefly reviewed. Then, the influence of residual errors is examined and different error concealment techniques are analyzed. Finally simulation results are presented.

8.2 TRANSMISSION SCHEME (MPEG-2 AND DVB)

8.2.1 MPEG-2 Video

The MPEG-2 international standard is very popular because it provides flexibility for various applications and is efficient for data compression. The Moving Pictures Expert Group (MPEG) was formed in 1988 to establish standards for various applications such as digital storage media, distribution, and communication.

The MPEG-2 video international standard [25] specifies the coded representation of the video data for broadcast-applications. The video source coding is based on motion compensated hybrid-DCT (discrete cosine transform)-coding. The hybrid-DCT-coding scheme is given in Figure 8.1. One distinguishes between two different picture coding types: *intra* and *inter* coding. Intra coded pictures (I-pictures) use only information from themselves (no motion compensation). Here, the block-based DCT is performed followed by quantization, zig-zag-scan, and variable length coding (VLC). The inter coded pictures use additionally motion compensated prediction from a previous and/or a next reference picture. Besides the motion vectors, the prediction error, which is coded like intra coded pictures, is sent in the bitstream. The predictive coded pictures (P-pictures) use only the previous reference picture for prediction. In addition, the bidirectionally predictive coded pictures (B-pictures) use the previous and the next reference pictures for prediction. Reference pictures are the I- and the P-pictures. I-pictures are also used as access points in the bitstream. Several pictures are grouped together in a group of pictures (GOP) which is characterized by the period of the reference pictures.

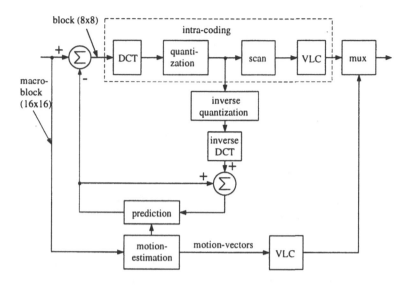

Figure 8.1   Hybrid-DCT-coder.

The pictures are divided into slices. One slice consists of several macroblocks (MBs). One MB is divided into four blocks, where each block is 8x8 pixels. This block is the entry for the 8x8 DCT.

The bitstream consists mainly of a sequence-header, a sequence-extension, a group-of-pictures-header, a picture-header, a picture-coding-extension, the picture data, and a sequence-end-code. The picture data is composed of slices, where each slice contains a slice-start-code and the data for each MB (Figure 8.2). Synchronization is possible at each slice due to the slice-start-code. The data for each MB, which is of variable length, starts with the MB address – coded as an increment – and after the motion vectors the DCT-coefficients are transmitted for each block. The codeword coded-block-pattern before the block-data indicates which of the six blocks (four blocks for the luminance component and two blocks for the chrominance components due to the 4:2:0 format) is coded or not. A block may not be coded, if for instance the prediction error is very small or if a MB is skipped, where the MB will be copied from the previous picture. Since in interlaced pictures frame and field motion vectors can be used, there exist up to two motion vectors for each MB in P-pictures and up to four in B-pictures (up to two in each direction). In inter coded pictures there could also be intra coded MBs. Additionally, MBs in B-pictures can also be predicted in only one direction.

In this chapter a TV picture of 4:2:0 format with 720x576 pixels is considered. 10 GOPs are simulated, where each GOP contains 12 pictures and the period of the

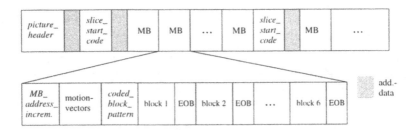

Figure 8.2   Bitstream-syntax for a picture.

reference pictures is 3. One slice consists of 44 MBs, so that one slice is one whole horizontal row of MBs in a picture. In this case there are 36 slices in a picture. The choice of the length of slices may have an influence on error concealment because synchronization is only possible at the beginning of a slice. Therefore, shorter slices could be better, but they will cause a higher bit rate. The bit rate of 5 Mbit/s is chosen, because it seems to be acceptable for a good TV quality (such as PAL or SECAM).

8.2.2   MPEG-2 Multiplexer

For error prone environments the MPEG-2 systems international standard [24] proposes the use of Transport Streams. In a Transport Stream the variable length bitstream, which comes from the video and audio source encoders, is converted to a constant length stream. The Transport Stream divides the VLC-data into fixed length packets of 188 bytes. Beside other necessary information like synchronization information, the 4 header bytes contain a transport-error-indicator, which is a one bit flag, which indicates if in this transport packet errors exist or if this packet is lost. This flag is important information for error concealment. It indicates which packet is erroneous and which is not.

8.2.3   DVB Transmission Scheme

For digital transmission of TV-signals in Europe, the DVB (digital video broadcasting) project submitted transmission systems for different transmission media (terrestrial, satellite, and cable) to the European Telecommunications Standards Institute (ETSI) [9, 10, 11]. Figure 8.3 shows the general transmitter for the DVB-transmission schemes DVB-T (terrestrial), DVB-S (satellite), and DVB-C (cable). The three systems have similar components to be as consistent as possible. Table 8.1 gives additionally an overview of the different components.

The transmission system is as follows: After multiplexing the compressed audio-, video-, and auxiliary-data in a Transport Stream, and after an energy dispersal unit,

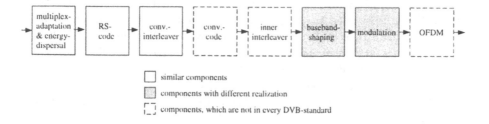

□ similar components

▨ components with different realization

⌐⌐ components, which are not in every DVB-standard

Figure 8.3   Components of the DVB-transmitter.

Table 8.1   Overview of the components of the DVB-standards.

	DVB-T	DVB-S	DVB-C
multiplex & energy-dispersal	MPEG-2, pseudo-random	MPEG-2, pseudo-random	MPEG-2, pseudo-random
RS-Code	yes	yes	yes
conv.-interleaver	yes	yes	yes
conv.-code	yes	yes	–
inner interleaver	bit-, frequency-interleaver,	–	–
baseband-shaping	–	root-cosine, roll-off: 0,35	root-cosine, roll-off: 0,15
modulation	QPSK, 16-, 64-QAM	QPSK	16-, 32-, 64-QAM
mapping	Gray	Gray	Gray
OFDM	yes	–	–

the packets are protected against channel errors with a channel encoder. Then, after baseband-shaping, modulation with Gray-mapping is applied and in the case of the critical terrestrial transmission OFDM (orthogonal frequency division multiplexing) is applied in combination with the inner frequency interleaver to combat the frequency selectivity of the channel. The receiver comprises the inverse components.

In both the satellite and the terrestrial transmission scheme, the channel encoder is based on a concatenated coding scheme with an outer Reed-Solomon(RS) RS-code, a convolutional interleaver with interleaver depth $I = 12$, and an inner convolutional code. The transmission over the less critical cable-channel is protected only with the outer RS-code. The outer RS-code is adapted to the transport packets of length 188 bytes. The outer code is a RS $(204,188,17)$ code over $GF(2^8)$, which can handle up to $t_{max} = 8$ errors. If there are more than t_{max} errors in a packet [6], the outer decoder can be used for error detection, except in the case in which the codeword is mapped into another codeword. However, the probability that one RS-codeword is mapped into another RS-codeword is very small. Therefore, the RS-decoder can detect packet errors with high reliability.

8.3 INFLUENCE OF RESIDUAL ERRORS ON THE VIDEO BITSTREAM AND ERROR DETECTION TECHNIQUES

After the RS-decoder at the receiver side, the resulting errors are burst packet errors. The inner decoder is a Viterbi-decoder, which will generate burst errors due to the fact that with too many errors the Viterbi decoder will choose a wrong path in the trellis diagram and will stay in this path due to the memory of the code. If the burst errors of the inner decoder are longer than the length of the interleaver depth, under bad reception conditions, the outer RS-decoder will cause burst packet errors. The packets may be erroneous or lost. In the first case one can try to make use of the erroneous data within a packet. In the second case, which happens if for instance the Transport Header is disturbed, the errors due to the lost data have to be concealed.

Errors can occur in the header codes or in the data. Errors in header codes may result in the loss of a picture, a group of pictures (GOP), or a sequence. But the percentage of the header codes is very small (0.14% in a sequence, where the sequence-header is sent every second GOP), so that the probability of the occurrence of a packet error in a header code is very small. Errors in the data are less critical. They result in a loss of a part of a slice or a whole slice. If errors are detected, they can be concealed. Un-detected errors will be visible in the picture domain as a wrong color or pattern (errors in DCT-coefficients), as a incorrectly predicted MB (errors in the motion vectors), or as MBs with wrong positions (errors in the MB address). Due to VLC, synchronization may be lost and errors will propagate. Since some of the parameters are differentially coded (e.g. the DC-value, the MB address, and the motion vectors), errors in those codewords will also propagate in the spatial domain. Additionally, errors will propagate in the time domain due to prediction. This is the case for errors in the reference pictures (I and P). Since B-pictures are not used as reference pictures, errors in B-pictures do not propagate in the time domain. When errors start to occur due to propagation it is better to stop the decoding process in order to avoid further impairments. Therefore, it is crucial to apply a good error detection technique.

The amount of the lost data depends mainly on the location of the error. If errors reside at the beginning of a slice, the remaining MBs of that slice will be lost even if the next packet contains error-free data of the same slice. This is due to the fact that synchronization is lost until the next slice in the case of errors. But in the case that errors are at the end of a slice, the length of the packet detrmines if the next slice is also affected. Table 8.2 gives the average number of packets in a slice and the number of MBs in a packet for the simulated sequence *Flower*. They depend on the picture coding type.

Table 8.2   Amount of packets in a slice (macroblock=MB).

	No. packets/ slice	No. MBs/ packet	
I-pictures	12.3	3.6	
P-pictures	5.5	8	
B-pictures	1.9	23	

To avoid errors in the picture domain, a good error detection technique should be applied. In the next Sections two different possible detection techniques are described and compared.

Error detection in the source decoder.   An obvious method for error detection is the use of the error detection capability of the source decoder. There are instances when the source decoder can detect the transmissions errors. Here are some:

- The VLC codeword cannot be mapped into the VLC table,

- undefined or reserved values are decoded for predefined values,

- wrong motion vectors appear,

- more than 64 DCT-coefficients are decoded,

- a combination of different codewords cannot occur in the bitstream (semantic), and

- header/synchronization codewords contain errors.

Since the source decoder is not designed for error detection, unfortunately, some errors will remain undetected. Those errors are for instance:

- A codeword is mapped into another codeword; If this new decoded word has a different length from the correct word because of the variable length coding, this error will propagate,

- errors in fixed length codewords, and

■   errors in header codewords, where the header will result in a different header.

To classify the error detection capability of the source decoder, the codewords were mapped into four classes. These were classes where error detection is: a.) always, b.) conditionally, c.) rarely, and d.) never possible. Codewords such as start codes, where nearly all errors will be detected, fall within the first class. The second class contains codewords like VLC-codewords, where error detection is possible but not always guaranteed. The third class includes codewords which contain reserved values, where error detection is possible only if those reserved values occur. Finally, the fourth class contains fixed length codewords, where no error detection is possible. The percentage of bits of the respective class of the simulated sequence *Flower* is given in Table 8.3. One can see that in about 26% of the data, errors will not/rarely be detected, in 68% of the data errors might be detected, and in only 6% of the data, errors will always be detected.

Table 8.3   Percentage of data in the four detection classes.

	error detection class			
	always	conditionally	rarely	never
$\sum$ Bits [%]	5.8	68.4	3.4	22.4

Figure 8.4 shows the results of error detection in the source decoder for the first I-picture of the sequence *Flower* disturbed with a packet error rate of $PER = 10^{-1}$. One can see that just before errors were detected visible impairments occur with this technique. After error detection the decoder stops and searches the next slice-start-code. The lost MBs are marked with black. Since visible impairments occur even with error concealment using the remaining lost MBs, better error detection is necessary.

Error detection in the channel decoder.   In addition to the error detection capabilities of the coder, one can also use of the error detection capability of the channel decoder. The channel decoder can detect errors in the received packets. An error-flag for each packet may be sent to the source decoder, which indicates if the packet contains errors or not. The outer RS-decoder detects un-decodable packets (containing errors) with very high reliability.

If the decoder knows that the packet contains errors, the data within the tansmitted data will not be used until the next synchronisation point (the next slice). This error detection technique finds the region where errors are in the bitstream. The region is 184 bytes long (without the Transport Header) and these 184 bytes will cause loss of synchronization until the next slice. This detection technique gives no information about the exact location of the errors in the packet. Therefore, in addition to the error-

Figure 8.4   I-picture, error detection in the source decoder, $\text{PER} = 10^{-1}$.

flag the reliability values of each symbol within the packets, which come from an inner Soft-Output-Viterbi-Algorithm (SOVA) [15], have to be used. With this information the data within the erroneous packet can be decoded until the first error-symbol [4]. Figure 8.5 shows the error detection with the RS-decoder and Figure 8.6 shows the error detection with the additional use of reliability values. One can see that additional 6 MBs can be recovered with the additional use of reliability values.

Figure 8.5   I-picture, error detection in the channel decoder, $\text{PER} = 10^{-1}$.

Figure 8.6   I-picture, error detection in the channel decoder with reliability information, $\text{PER} = 10^{-1}$.

8.4 ERROR CONCEALMENT TECHNIQUES

8.4.1 State of the art

After the finalization of the JPEG and H.261 standards in 1990s a lot of work was devoted to the error concealment problem for block-based coding techniques. Two different types of error concealment exist: spatial response frequency error concealment, which makes use of the spatial redundancy within a picture, and temporal error concealment, where the temporal redundancy between consecutive pictures of a video sequence is used. In the following the state of the art in error concealment techniques is reviewed.

Spatial error concealment. In [37, 38] a technique for reproducing lost DCT coefficients is proposed. Here, the missing coefficients are estimated by applying a maximal smoothness constraint at the border pixel of the missing block. In [41, 42] this technique is extended for reproducing the lost DCT-coefficients in inter coded pictures. Furthermore, the smoothness measure is enhanced by using the second order derivative in [40]. A similar technique computed in the pixel domain is proposed in [17]. An interpolation of the pixel values from 4 corner pixels is suggested in [14]. In [26] a comparison between a simple interpolation of the first five DCT-coefficient with the technique proposed in [37] is given. It is shown that the technique proposed in [37] is only better for the first 5 DCT-coefficients than interpolation. For higher DCT-coefficients above some frequency level, simply setting the coefficients to zero is as effective as or more effective than any other technique. Another technique is proposed in [30], which is a similar technique to that proposed by the author, where the pixel values are obtained by a weighted interpolation of the neighboring pixels. In addition, a maximum a posteriori (MAP) estimation is proposed in [31], where the image content is modelled as a Markov random field. A comparison in [28] showed that a proposed multi-dimensional non-linear filter is better than applying MAP-estimation.

The main problem for all the proposed techniques is their poor performance in reproducing high spatial detail. In MPEG-2 coded sequences the lost area is additionally very large, which makes it very difficult to reproduce lost image content with high spatial detail. The emulation of edge plays an important role in the fields of image restoration, image enhancement, and postprocessing [27]. In [27] a directional filtering scheme for postprocessing is used. The experience of this field can also be used for error concealment. In [32, 23, 18, 33] the directional filtering is applied in different ways. Since missing blocks may contain more than one edge direction, the directional interpolation can also be applied in more directions as suggested in [19, 39].

Temporal error concealment. Many publications in the field of temporal error concealment can be found in the literature. Besides techniques, where the motion vectors are sufficiently protected, and the simple copying technique of the previous picture to the location of the missing blocks, there are many proposals, where missing motion vectors are estimated from neighboring MBs [1, 21, 22, 12] and the missing blocks are predicted in a motion compensated way. Additionally the average value [36, 29, 26, 22] or the Median value [16, 26, 22] of the neighboring motion vectors may be considered, where the Median value provides better results than the average value [26]. The new motion vector can be chosen by computing the minimum mean square error (MMSE) of the boundary pixels as suggested in [22, 12]. The motion vectors of the previous picture can also be considered [16, 26, 22, 12]. Finally, the prediction technique should be chosen properly as proposed in [29].

Further techniques. The motion compensated temporal error concealment techniques promise very good results in moving pictures with uniform motion. However, in pictures with highly irregular motion or scene changes, spatial error concealment techniques may give better results. Therefore, the combination of both techniques may be considered [34].

Additional improvements will be obtained by transmitting more synchronization words in the bitstream (more slices). Furthermore with advance synchronization before the next slice one can enhance the final picture quality [20, 7, 13]. In the following different error concealment techniques investigated in this chapter are analyzed.

8.4.2 Temporal error concealment

Copying. The simplest method for error concealment is copying from the previous picture. The previous picture will be a reference picture (I or P-picture). First, the whole previous reference picture may be copied, but this will cause visible effects in the following pictures due to the prediction. Figure 8.7 shows a P-picture after a previous P-picture, which was concealed with the whole reference picture before. The motion vectors use wrong reference MBs, therefore, the quality will be reduced. Also B-pictures will suffer from copying the whole content in a reference picture. Figure 8.8 shows the results for a B-picture before the concealed P-picture. Hence, the simplest solution for handling errors in reference pictures is to just stop the sequence for the whole GOP. However, errors in B-pictures will not temporally propagate. Therefore, it is possible to conceal B-pictures with the whole content of the previous picture, where the sequence will only be stopped for one picture.

Secondly, one can use all the correct received data and copy only the areas where the packet errors are. This will result in a better quality, since only the missing MBs will be replaced. But still there are visible shifts, if there is motion in a sequence. Therefore, it is proposed to use motion compensated temporal error concealment as described in the next Section. Figure 8.9 shows a P-picture disturbed with a packet error rate of $PER = 10^{-1}$ and Figure 8.10 shows the copying method. The errors are

Figure 8.7 P-picture after copying the whole previous P-picture.

Figure 8.8 B-picture after copying the next P-picture.

detected with the channel decoder with the additional use of reliability values (perfect symbol-error detection is assumed).

Figure 8.9 P-picture, $PER = 10^{-1}$.

Figure 8.10 P-picture, copying.

Motion compensated temporal error concealment (P-/B-pictures). In addition, the motion compensated temporal error concealment (MC-TEC) uses the motion vectors from neighboring MBs in the current picture (if motion vectors exist). There are several methods for the motion vector selection. First, the motion vectors from one of the nearest MBs may be used [2]. The second technique takes into account the motion vectors of all the surrounding (existing) MBs and incorporates the lost MBs with the possible motion vectors (also the Median motion vectors and no motion vectors may be considered). The best matching MB is chosen following the minimum mean square error (MMSE) criterion of the border pixels for each field. The technique (frame/field, which field is used) for incorporating the MB with the Median value is selected from the technique of the surrounding MBs, where the criterion for the chosen technique is again the MMSE of the border pixels. This is applied iteratively, with no motion vectors being considered for the last iteration. Simulation results in terms of PSNR-values showed that both techniques give similar results. If one uses field and frame vectors, the results depend strongly on the prediction technique (frame/field, which field is used), therefore, the additional use of the Median value gives no additional improvements [5].

Figure 8.11 shows the MC-TEC technique (motion vector selection with the MMSE criterion) with error detection in the channel decoder with the use of reliability values for the disturbed P-picture.

Figure 8.11   P-picture, MC-TEC.

Temporal error concealment for I-pictures.   The MC-TEC technique, which uses the motion vectors of the neighboring MBs, performs very well. Since motion vectors are available only in P and B-pictures, it may only be used for them. Because I-pictures have a higher data rate, it is more likely that I-pictures contain errors. Therefore, it is crucial to have a good error concealment technique for them. For I-pictures, MPEG-2 has the possibility of transmitting additional error concealment motion vectors in the bitstream. This would result in a noticeably better quality and is not as difficult to implement as the spatial concealment algorithms or combinations of spatial and temporal concealment techniques. With the additional transmission of frame motion vectors the overhead will be less than 0.7% of the total bit rate [35]. But sending of additional motion vectors is only possible if the transmitter is able to do this additional motion estimation. This will result in higher costs at the transmitter side and it will

248

depend on the provider. Since it is only an option, here, it is assumed that no additional concealment vectors are transmitted.

To make use of the advantages of the MC-TEC technique, the motion vectors of neighboring pictures may also be used for I-pictures. The temporal correlation between pictures of a sequence is very high in the motion direction. In a sequence with uniformly motion, the motion vectors of the previous picture can be used for error concealment. However, if an object moves very fast, the motion vectors of the previous picture may not be appropriate. The position of the object in the current picture may be in another position than in the previous picture. Therefore, a technique is proposed, which moves the motion vectors along the motion direction. Here, the motion vectors of the previous reference picture are stored, displaced, and used for error concealment. The motion vectors of the previous B-pictures (in time) cannot be used, since they use the I-picture for prediction and one needs to decode another picture, which depends on the I-picture. For decoding of the next picture, the error concealed I-picture is needed. Therefore, the nearest motion vectors which can be used are the motion vectors of the previous reference picture (I or P).

The motion vectors of the previous picture are displaced along the motion direction. This can be done by moving the virtual center of a MB corresponding to the motion vectors. If the shifted center lies in the current MB, the motion vectors will not be displaced. Otherwise, the motion vectors belong to the MB where the shifted center lies. Figure 8.12 shows the displacement of motion vectors.

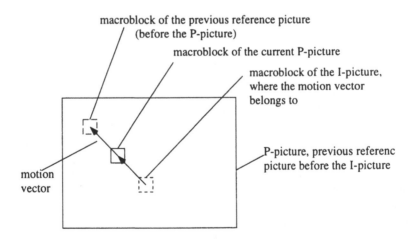

Figure 8.12   Displacement of the motion vector.

Assume the motion vectors $v(x, y, r, s, t), r, s, t = 0, 1$ of a picture with the ITU-R 601 format, where (x, y) is the spatial MB position, r is the index of the field/frame vector(s), s is the index for the prediction direction (forward/backward), and t is the index for the spatial component of the motion vector. Since only motion vectors of P-pictures are used, the motion vectors are independent from s: $v(x, y, r, t), r, t = 0, 1$. Therefore, for each MB position there exist up to two motion vectors with two spatial components. In the case of field prediction the average of the motion vectors is computed as:

$$v_m(x, y, t) = (v(x, y, 0, t) + v(x, y, 1, t))/2 \quad \text{for} \quad t = 0, 1 \quad . \tag{8.1}$$

With this, the motion vectors $v_m(x, y, t)$ depend only on the MB position and on the index t. The new position (x_p, y_p) of the motion vectors is computed as:

$$x_p = (((16 \cdot x) + 8) - v_m(x, y, 0)) \bmod (16) \quad , \tag{8.2}$$

$$y_p = (((16 \cdot y) + 8) - v_m(x, y, 1)) \bmod (16) \quad . \tag{8.3}$$

The motion vector(s) $v(x, y, r, t)$ will be moved to the position (x_p, y_p):

$$v(x_p, y_p, r, t) = v(x, y, r, t) \quad \text{for} \quad t = 0, 1, r = 0, 1 \quad . \tag{8.4}$$

This will be done for all MB positions. The corresponding prediction technique will also be moved with the motion vectors. After completing of the motion vector displacement algorithm, one obtains a motion vector field with the displaced vectors. If more than one motion vector correspond to a MB, the motion vector which yields the best matching MB (MMSE criterion for the border pixels computed for each field) is used for error concealment. For those MBs which in the end have no motion vector available, the motion vectors of the surrounding MBs and the un-shifted motion vectors of the current MB are investigated and the motion vectors with MMSE are chosen. In most cases, those MBs contain uncovered MBs. Therefore, uncovered MBs are detected with this algorithm and can also be concealed with another error concealment algorithm (for instance spatial error concealment).

In Figure 8.6 one I-picture without error concealment disturbed with a packet error rate of PER $= 10^{-1}$ was shown. The results for copying, MC-TEC without displacement of the motion vectors, and MC-TEC with displacement of the motion vectors are given in Figure 8.13-8.15, respectively. One can see that errors in the area of the tree are concealed much better with the motion vector displacement. The regions where the MB contains no motion vector after displacement are given in Figure 8.16. These MBs are also uncovered MBs.

The displaced motion vectors of the previous reference picture may also be considered for error concealment in P and B-pictures. This will enhance the error concealment in MBs, where the motion of the neighboring MBs is different.

Figure 8.13 I-picture, copying.

Figure 8.14 I-picture, MC-TEC without displacement of motion vectors.

Figure 8.15 I-picture, MC-TEC with displacement of motion vectors.

Figure 8.16 I-picture, MC-TEC, MBs without motion vectors.

8.4.3 Spatial error concealment

Spatial error concealment techniques are devoted to I-pictures for which no motion information exists. Spatial error concealment techniques make use of the spatial redundancy in a picture. One method based on weighted interpolation is considered in this chapter [3]. This technique interpolates each pixel of the entire $2N \times 2N$ MB with the adjacent pixels of the four neighboring MBs. Figure 8.17 shows the MB with the boundary pixels of the neighboring MBs ($N = 4$).

Each pixel of the current MB with the size $2N \times 2N$ will be concealed by weighted interpolation of the four pixels of the surrounding MBs. The equation describing this

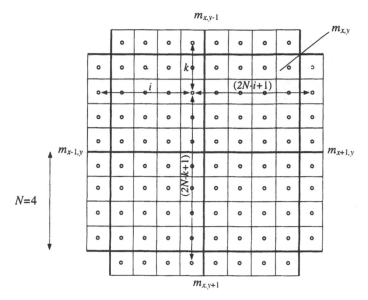

Figure 8.17   Spatial interpolation of the entire MB.

process is:

$$m_{x,y}(i,k) = \frac{1}{2(2N+1)} \quad ((2N - i + 1) \cdot m_{x-1,y}(2N, k) + i \cdot m_{x+1,y}(1, k) +$$
$$(2N - k + 1) \cdot m_{x,y-1}(i, 2N) + k \cdot m_{x,y+1}(i, 1)) \ (8.5)$$
$$\text{for} \quad i, k = 1 \dots 2N \quad ,$$

where $m_{x,y}(i, k)$ is the pixel value (i, k) of the MB at the spatial position (x, y). With this equation, a pixel value is obtained by an interpolation of the four pixel values of the neighboring MBs weighted with the opposite distance.

This concealment technique works very well if the surrounding MBs exist. Figure 8.19 shows the interpolation, where every fourth MB is disturbed (Figure 8.18). Here, a very non-realistic slice-length of one MB is considered. With the interpolation technique no abrupt transition occurs in a MB (no 'block-effect'). The entire MB becomes blurred if there are edges in a MB. This then will be a visible 'block- effect' in terms of MBs as well. High spatial detail like the flowers will not be reproduced correctly.

However, in MPEG-2 the synchronization points are at the start of the slices, therefore, in the case of errors horizontal stripes of MBs must be concealed. The left and the right MBs cannot be used for concealment because they do not exist. If some of the MBs do not exist, the corresponding opposite distance will be set to zero during interpolation (for instance, if $m_{x-1,y}$ does not exist $(2N - i + 1)$ will be set to zero).

Figure 8.18   I-picture, every 4th MB dis-   Figure 8.19   I-picture, spatial interpola-
turbed.   tion.

With only two available MBs, $m_{x,y-1}$ and $m_{x,y+1}$, Equation 8.5 becomes

$$m_{x,y}(i,k) \;=\; \tfrac{1}{2N+1} \left((2N - k + 1) \cdot m_{x,y-1}(i, 2N) + k \cdot m_{x,y+1}(i,1) \right) \quad (8.6)$$
$$\text{for} \quad i, k = 1 \ldots 2N.$$

This spatial interpolation can be used for MPEG-2 coded pictures with long slice-lengths. Figure 8.20 shows the spatial interpolation of the disturbed I-picture (Figure 8.6, PER $= 10^{-1}$).

Figure 8.20   I-picture, spatial interpolation, PER $= 10^{-1}$.

Since only the top and the bottom MB can be used for interpolation, the interpolation results in an interpolation in one direction, the vertical direction. Edges in the vertical direction will be interpolated correctly, whereas edges in the horizontal direction will become blurred. Also the problem of reproducing high spatial detail remains.

An extension of the spatial interpolation is the interpolation of one pixel with more pixel values of the surrounding MBs. One technique investigated here, is multidirectional weighted interpolation in eight directions. The eight directions are depicted in Figure 8.21.

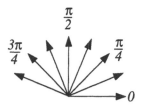

Figure 8.21   The eight interpolation directions.

In Equation 8.5 only two directions – the horizontal and vertical direction – are considered. The same principle is used now for all eight directions. In addition, in one direction more than only the border pixel values can be used for interpolation. For one direction, for instance the horizontal direction, the equation is:

$$m_{x,y}(i,k) = \frac{1}{\sum\limits_{j=1}^{2N} (w_{\mathrm{L}}(i,j) + w_{\mathrm{R}}(i,j))} \sum_{j=1}^{2N} (w_{\mathrm{L}}(i,j) \cdot m_{x-1,y}(j,k) +$$

$$w_{\mathrm{R}}(i,j) \cdot m_{x+1,y}(j,k)) \qquad (8.7)$$

with

$$\begin{aligned} w_{\mathrm{L}}(i,j) &= (2N - i + 1) + (j - 1) \quad , \\ w_{\mathrm{R}}(i,j) &= i + (2N - j) \quad , \quad \text{and} \quad i,k = 1\ldots 2N \quad . \end{aligned}$$

The weights w_{L} and w_{R} are the opposite distances of the pixel value $m_{x-1,y}(j,k)$ and $m_{x+1,y}(j,k)$ to the pixel value $m_{x,y}(i,k)$. This interpolation will be computed for all eight directions. It has been shown that using more than just the border pixels gives subjectively no better results than the use of the border pixel values only (for $j = 2N$ in Equation 8.7). This is due to the fact that pixel values which are very far from the interpolated pixel value and contain perhaps other image content, influence the interpolated value. Since the lost pixel area – at least a one MB area – is very large, the image content may change within the neighboring MBs. Also, investigations with the additional use of a spline-interpolation, where every second pixel is interpolated first and the remaining pixels are interpolated by spline-interpolation, failed.

Figure 8.22 shows the multi-directional interpolation (4 directions) using more than just the border pixels in each direction. One can see that the missing MBs become blurred and the smoothness criterion is not fulfilled anymore. Figure 8.23 shows the results for considering only the border pixels for the interpolation (8 directions), which is subjectively better than using more than just the border pixels.

Figure 8.22 I-picture, multi-directional interpolation with more than only the border pixels.

Figure 8.23 I-picture, multi-directional interpolation with only the border pixels.

With edges as an image content, the edges become blurred with the multi-directional interpolation. To avoid this, one can try to perform a directional filtering in that direction. The direction can be determined by computing the gradients. For every pixel value of the neighboring MBs $m_{x,y}$ the local components of the gradient can be computed as [8, 33]:

$$
\begin{aligned}
g_{\mathrm{x}} &= m_{x,y}(i+1,k-1) - m_{x,y}(i-1,k-1) + 2m_{x,y}(i+1,k) - \\
&\quad 2m_{x,y}(i-1,k) + m_{x,y}(i+1,k+1) - m_{x,y}(i-1,k+1) \quad, \\
g_{\mathrm{y}} &= m_{x,y}(i-1,k+1) - m_{x,y}(i-1,k-1) + 2m_{x,y}(i,k+1) - \\
&\quad 2m_{x,y}(i,k-1) + m_{x,y}(i+1,k+1) - m_{x,y}(i+1,k-1) \quad. \quad (8.8)
\end{aligned}
$$

This is equivalent to applying the 3×3 Sobel-operators:

$$
S_{\mathrm{x}} = \begin{bmatrix} -1 & 0 & 2 \\ -2 & 0 & 2 \\ -1 & 0 & 1 \end{bmatrix} \quad S_{\mathrm{y}} = \begin{bmatrix} 1 & 2 & 1 \\ 0 & 0 & 0 \\ -1 & -2 & -1 \end{bmatrix} \quad. \quad (8.9)
$$

The magnitude G and the edge orientation θ of the gradient will be computed as:

$$G_r = \sqrt{g_x^2 + g_y^2} \quad \text{and} \quad \theta = \arctan(g_y/g_x) \quad . \tag{8.10}$$

These two values will be computed for every pixel of the neighboring MBs for each field. The computed edge orientation will be classified into the eight directions by choosing the nearest direction. If the computed direction cuts the missing MB [33], the magnitude will be added to a counter of each direction. The resulting direction will be the direction with the highest counter. Additionally, if the counter lies below a threshold, multi-directional interpolation will be applied.

Figure 8.24 shows the results for the directed interpolation. Some MBs with edges will be reproduced better. However, the edges do not always match. In addition, the flowers will be reproduced with directed interpolation, where multi-directional interpolation gives subjectively better results. An adaptive technique, where the missing MBs with high spatial detail will be concealed with multi-directional interpolation in any case, will provide better results.

Figure 8.25 shows the results for multi-directional interpolation (8 directions) for the disturbed I-picture with packet errors. It has been shown that the gain for multi-directional interpolation is only 0.7 dB at $PER = 10^{-1}$. Therefore, in the simulations the less complex interpolation of Equation 8.6 is used for spatial error concealment. Altogether one can say that the subjective quality can be increased by applying multi-directional interpolation. Due to the fact that the missing area is very large, the directed interpolation gives no significant improvements.

Figure 8.24 I-picture, directed interpolation.

Figure 8.25 I-picture, multi-directional interpolation (only border pixels), $PER = 10^{-1}$.

8.4.4 Combination of spatial and temporal error concealment

The spatial and temporal error concealment techniques may be combined within a video sequence by applying spatial error concealment to I-pictures with scene changes or no concealment motion vectors and temporal error concealment to P and B-pictures. In addition, both techniques may be combined within an I-picture [34]. Here, copying is used for temporal error concealment under the assumption that no motion vectors are available. The criterion for the decision between spatial and temporal replacement is the measurement of the image activity. The motion will be computed as the mean square error (MSE) between the nearest neighboring MBs (top or bottom) of the current $(m_{x,y-1/x,y+1})$ and the previous picture $(m'_{x,y-1/x,y+1})$:

$$ \text{MSE} = \text{E}\{(m_{x,y-1/x,y+1} - m'_{x,y-1/x,y+1})^2\} \quad . $$

The spatial detail will be computed as the variance of the nearest neighboring MB (top or bottom) of the current picture:

$$ \text{VAR} = \text{E}\{m^2_{x,y-1/x,y+1}\} - \text{E}\{m_{x,y-1/x,y+1}\}^2 \quad . $$

In addition, a third criterion can be used for a more reliable decision [4]: The spatial detail of the previous picture is computed as:

$$ \text{VAR}' = \text{E}\{m'^2_{x,y-1/x,y+1}\} - \text{E}\{m'_{x,y-1/x,y+1}\}^2 \quad . $$

The decision rules are plotted in Figure 8.26. There are three thresholds used for the decision. In the area, where both techniques are plotted, the decision is based on the third threshold. If VAR' lies above the threshold, copying is applied, otherwise spatial error concealment is applied. A more detailed description can be found in [4].

Since both techniques – copying and spatial error concealment provide similar results in terms of PSNR-value, the combination of them is not better than both techniques. However, the subjective results are better than one of them alone (Figure 8.27). But still the offset of the motion will be visible in the area of the flowers.

8.4.5 Error concealment techniques for the MPEG-2 hierarchical profiles

Spatial Scalable Profile. The MPEG-2 spatial scalability provides the link between TV and HDTV resolution. The base layer bitstream contains all data for TV quality (MP@ML). The upper layer contains all additional data for obtaining HDTV resolution. In addition to the motion vectors, the upsampled TV picture is used for prediction and the prediction error is sent in the upper layer. In I-pictures the prediction is only made with the upsampled TV-picture. In P and B-pictures motion compensation is also performed.

Because the TV layer is the most highly protected layer, the base layer may be received without errors at bad reception conditions, but the upper layer may contain

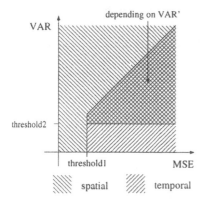

Figure 8.26 Decision rules.

Figure 8.27 I-picture, combination of copying and spatial error concealment.

packet errors. If one has a stationary receiver with HDTV quality, it seems worthwhile to conceal the corrupted packets of the upper layer instead of using only the base layer.

The best method of error concealment is to conceal the damaged area with the up-sampled TV-picture [3]. With this technique, the picture quality will slightly decrease (sharpness), but no shifts and wrong MBs will occur as is the case with copying or motion compensated temporal error concealment. This technique gives good results by applying it to both intra and inter coded pictures.

SNR Scalable Profile.   If there are errors in the upper layer computed by SNR-scalability, it will be better to take at the erroneous part of the image only the base layer information, because the upper layer contains only the quantization error (without motion information). If one tries to reproduce the quantization error, it may happen that the estimated error is wrong and this incorrectly estimated error will result in a low picture quality.

8.4.6   Additional techniques which capitalize on the remaining correctly received data

Re-synchronization.   After the occurrence of a wrong symbol, the decoder can try to re-synchronize *before* the next slice and use the correct received information. For re-synchronization at MB level, the following algorithm will be applied [20]: The decoder will start to decode one MB at the current position in the bitstream. If the MB is not detected as a wrong MB from the error detection in the source/channel decoder, the re-synchronization is assumed to be correct and the decoder will continue either until the next slice-start-code is received, or until another error is detected by error detection in the source/channel decoder. If the MB will be declared as a wrong MB, the decoder will repeat decoding a MB by starting one bit after the last position. This process will be repeated until a correct received MB can be found or until the next slice-start-code is received. Figure 8.28 shows the flowchart of the re-synchronization process.

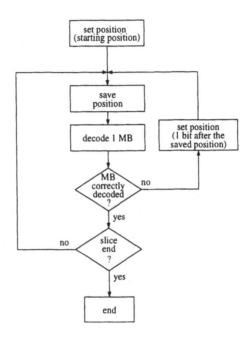

Figure 8.28   Flowchart for the re-synchronization process.

Here, the correct received MBs will be grouped into subslices, where each subslice contains all MBs which are correctly decoded consecutively after re-synchronization. The subslice ends if there is a slice-start-code received. It also ends if there is another symbol error or if there is a MB not correctly decoded. If the subslice contains only one MB, the search is started one bit after the beginning of the MB to improve the re-synchronization process, since it is most likely, that this MB is not correctly re-synchronized. The subslice having only one MB will be discarded.

Additionally, since the first few MBs of a subslice may not be re-synchronized correctly (for instance wrong DC- and AC-values), the subslices in I-pictures are further divided before positioning, if neighboring MBs of one subslice do not fit. The criterion for the smoothness is the MSE between the border pixels of two neighboring MBs.

However, especially in I-pictures, there is much information which is lost if one decodes a MB without having decoded the previous MB. The following information has to be looked at carefully: i) the *quantizer-scale-factor*, ii) the *motion vectors* (P-/B-pictures), iii) the *DC-value* (I-pictures), and iv) the *MB-address*. The concealment of these parameters will be briefly described in the following. A more detailed description can be found in [5].

i) **Quantizer-scale-factor:** The previous decoded quantizer-scale-factor is used to prevent a *really incorrect* quantizer-scale-factor.

ii) **Motion vectors (P-/B-pictures):** The differentially coded motion vectors in P-/B-pictures of the first MB of a subslice will be concealed with the motion vectors of the nearest neighboring MBs (top and/or bottom).

iii) **DC-value (I-pictures):** The DC-value in I-pictures is also differentially coded. Different techniques may be considered for concealing the first MB of a subslice [20, 7]. In this chapter the mean DC-value of the neighboring MBs is subtracted from the mean DC-value of the re-synchronized subslice, and this correction factor is simply added to the pixel values of the subslice [5]. With this additional technique wrong DC-value propagation is compensated and as many re-synchronized MBs as possible are used within the subslice. In addition the positioning technique will be more reliable with this DC-correction.

iv) **MB address:** The MB address is coded in increments so that the correct location of the MB will not be known. Therefore, a positioning technique has to be applied. The best location for the subslice is the location with MMSE of the border pixels. Here, long subslices are located before short subslices to avoid the case that short subslices limit the positions of long subslices. The best location is computed with the corrected DC-value.

At the end of the positioning, there will be several subslices available which contain only a few MBs which are sometimes incorrect (e.g. wrong AC-coefficients) due to wrong re-synchronization, that should be deleted. The criterion for deleting them is a threshold MSE of the border pixels of the MB. The deleted MBs have to be further concealed. The concealment techniques described in the previous Sections can be used for error concealment for the remaining lost MBs.

To improve the concealment techniques in regions where more than one slice is disturbed, the algorithms can be applied iteratively. For re-synchronization subslices, which initially have no neighboring MBs, can then be positioned and concealed, if the neighboring subslices are already concealed.

Figure 8.29 shows the re-synchronized subslices of the disturbed I-picture and Figure 8.30 shows re-synchronization with MC-TEC of the remaining lost MBs.

The use of short slices. The use of a shorter slice-length allows one to synchronize faster. However, the data rate will increase or the picture quality will decrease. Since packets contain different numbers of MBs in the different types of pictures, it seems

Figure 8.29 I-picture, re-synchronized subslices.

Figure 8.30 I-picture, re-synchronized subslices and concealment of the remaining lost MBs.

worthwhile to adapt the slice-length to the picture coding type. For instance packets in B-pictures contain 23 MBs on average. Therefore, it makes no sense to use a slice-length which is shorter than 22 MBs. The average number of MBs in a packet was given in Table 8.2. The use of a short slice-length mainly in I-pictures will improve the final picture quality considerably at low cost of complexity and data rate.

8.5 SIMULATION RESULTS

The simulations were carried out under the following conditions: The bit rate of the MPEG-2 MP@ML bitstream was 5 Mbit/s. I-, P-, and B-pictures of different sequences were disturbed with a packet error rate of $PER = 10^{-1}$, $PER = 2 \cdot 10^{-2}$, and $PER = 10^{-2}$ ($BER = 2.3 \cdot 10^{-3}$, $BER = 4.4 \cdot 10^{-4}$, and $BER = 2.4 \cdot 10^{-4}$, respectively) with twenty different error patterns. The mean-value of the peak-signal-to-noise-ratio (PSNR) for the luminance component (Y) is given. Burst errors were simulated with the *digital channel model*, which reproduces the errors after the channel decoder (Rayleigh-channel) [2].

For I-pictures of the sequence *Flower* different error concealment techniques are investigated. The mean PSNR-values for copying, spatial error concealment (SEC), and MC-TEC with displacement of motion vectors using error detection in the source and the channel decoder with and without reliability values (SOVA/RS) are depicted in Figure 8.31 (PSNR without errors: 31.4 dB). In addition, the results for re-synchronization and for the use of shorter slice-lengths are given in this Figure.

The simulation results show that the PSNR-value is up to 4.6 dB higher for MC-TEC compared to copying. The gain for the additional use of re-synchronization is

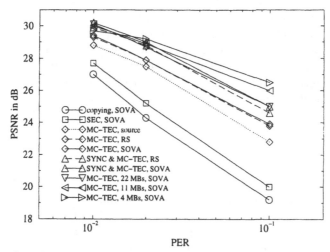

Figure 8.31   PSNR-comparison between different error concealment and error detection techniques (I-pictures).

up to 1.1 dB. The difference between the different detection techniques is shown in combination with MC-TEC. One can see that error detection in the channel decoder in combination with a good error concealment technique outperforms error detection in the source decoder ('source'). With the additional use of reliability information (perfect symbol error detection is assumed) a gain can be only obtained in combination with re-synchronization and a shorter slice-length. The PSNR-values of the combination of copying and SEC within a picture – not depicted in the Figure – lie between both techniques, but it will be interesting to investigate the improvements with subjective tests.

Figure 8.31 gives also the PSNR-values for the disturbed I-pictures, where the slice-length is shorter (22, 11, and 4 MBs instead of 44 MBs per slice (PSNR without errors: 31.4 dB, 31.0 dB, and 30.5 dB)). The simulation results show that the gain when using shorter slices (4 MBs per slice) in combination with MC-TEC is 7.3 dB at PER $= 10^{-1}$ compared to copying without the use of shorter slices. With these additional synchronization words in the bitstream, the picture quality can be increased in the case of errors without applying a complex algorithm. However the final picture quality will be slightly reduced. In the simulations a constant slice-length for all types of pictures is considered. If an adapted slice-length is considered, the final picture quality will be better.

The results show that MC-TEC in combination with the use of a shorter slice-length and error detection in the channel decoder outperforms all other techniques. With a given long slice-length, MC-TEC should be applied in I-pictures with no scene changes and irregular high motion, where a scene change/motion detector may be helpful for the decision, which concealment technique should be applied.

Simulation results for copying and MC-TEC, where the best matching motion vectors are chosen, in combination with different error detection methods in P-pictures are given in Figure 8.32 for the sequence *Flower* (PSNR without errors: 31.4 dB). The results show that a gain of 4.8 dB can be achieved with MC-TEC. The gain for error detection with the RS-decoder is 1.5 dB. Since MC-TEC performs very well, the additional use of reliability values (SOVA) gives only small improvements. In Figure 8.32 the re-synchronization is depicted likewise. The main problem of re-synchronization in inter coded pictures is that incorrect re-synchronization may cause error propagation for the whole subslice (e.g. a wrong motion vector difference). Therefore, no additional gain can be achieved with this technique. In addition, the results for shorter slice-lengths are given in Figure 8.32 (22, 11, and 4 MBs instead of 44 MBs per slice (PSNR without errors: 31.4 dB, 31.3 dB, and 31.1 dB)). A gain of up to 7.4 dB can be achieved with a slice-length of 4 MBs and the use of MC-TEC compared to copying without the use of shorter slices. The main advantage of the use of shorter slices with an accurate error positioning technique is the possibility of faster re-synchronization without concealing the differentially coded values.

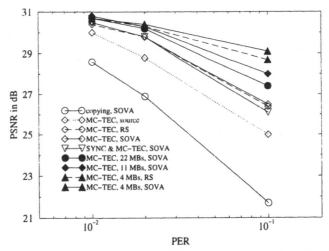

Figure 8.32   PSNR-comparison between different error concealment and error detection techniques (P-pictures).

Simulation results for B-pictures are given in Figure 8.33 (PSNR without errors: 30.5 dB). The results are similar to the results for P-pictures. In B-pictures fewer slices are erroneous at the same PER since B-pictures contain less data. The results show that MC-TEC outperforms all techniques and that the reduced picture quality when using shorter slices (22, 11, and 4 MBs instead of 44 MBs per slice (PSNR without errors: 30.5 dB, 30.3 dB, and 29.9 dB)) dominates the results. One can see that with a slice-length shorter than 22 no additional gain can be achieved.

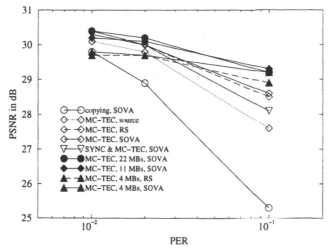

Figure 8.33   PSNR-comparison between different error concealment and error detection techniques (B-pictures).

The results are carried out under the condition that errors are only within one picture. To show the effects of errors within the whole sequence, the sequence *Flower* was disturbed with PER $= 10^{-2}$ and the PSNR values are depicted in Figure 8.34 for each picture. Different techniques for I- resp. P-/B-pictures are combined. One can see that MC-TEC outperforms the other techniques. The use of re-synchronization in I-pictures in most cases is better than MC-TEC, but with wrong re-synchronized subslices the picture quality can be decreased as it is the case with picture no. 61.

The PSNR-values for different error concealment techniques in P and B-pictures for the hierarchical MPEG-2 spatial scalable profile are given in Table 8.4 for the sequence *Ski*. The enhancement layer of an MPEG-2 scalable coded sequence with a bit rate of $6 + 9 = 15$ Mbit/s is achieved with spatial scalability. This layer was disturbed with a PER $= 10^{-1}$. Error concealment is done with the upsampled TV-picture (U-TV). For comparison, the PSNR-value is also given for both layers without errors and for the base layer only.

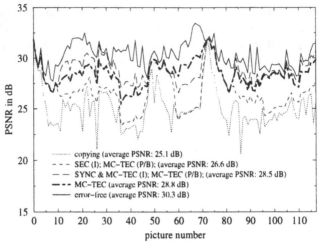

Figure 8.34   PSNR-comparison for a disturbed sequence (PER $= 10^{-2}$).

Table 8.4   PSNR comparison, disturbed enhancement layer.

	PSNR values for *Ski*			PSNR values for *Flower*		
	base layer	U-TV	without errors	only low. layer	err. enh. layer	without errors
P: Y	24.8	31.1	32.2	26.6	28.8	29.0
B: Y	24.8	30.4	31.3	25.0	26.7	26.8

The gain for the use of U-TV is 6.3 dB. The results show that it is better to conceal the wrong parts of a picture with the technique U-TV than to use the base layer only. For I-pictures the results will be similar.

The PSNR-values for picture replenishment in P and B-pictures for the hierarchical MPEG-2 SNR scalable profile are also given in Table 8.4 for the sequence *Flower*. The enhancement layer of an MPEG-2 scalable coded sequence with a bit rate of $3 + 3 = 6$ Mbit/s is achieved with SNR scalability. This layer was disturbed with PER $= 10^{-1}$. The technique for picture replenishment consists of using only the base layer where errors were detected. For comparison the PSNR-value is given for the two layers without errors and for the base layer only (the enhancement layer is not decoded). The results show that it is more worthwhile to also use the correct received information of the disturbed enhancement layer than to use the lower layer only.

8.6 CONCLUSIONS

In this chapter error concealment techniques which use enhanced error detection methods for MPEG-2 video sequences applied to digital TV/HDTV broadcasting are studied. These techniques are based on temporal, spatial, frequency error concealment, and the combinations thereof. In addition, a re-synchronization technique is also investigated where the correctly received data are utilized as much as possible. It has been shown, that the use of error detection in the channel decoder helps to avoid incorrectly decoded macroblocks.

The simulation results show that even if residual errors enter the bitstream the picture quality can be improved by applying post processing error concealment techniques in the decoder. Generally, the more complex the algorithm for error concealment the better is the picture quality. However, some relatively simple algorithms such as motion compensated temporal error concealment yield also very good results. For a scalable codec it was shown that it is better, instead of discarding the entire enhancement layer, to use the correctly received information and conceal the rest when reconstructing the coded image.

References

[1] G. Aartsen, R.H.J.M. Plompen, and D.E. Boekee. Error resilience of a video codec for low bitrates. In *Proceedings IEEE International Conference on Acoustics, Speech, and Signal Processing ICASSP'88*, New York, USA. April 1988, pp. 1312-1315.

[2] S. Aign and K. Fazel. Error detection & concealment measures in MPEG-2 video decoder. In *Proceedings International Workshop on HDTV'94*, Turin, Italy, October 1994.

[3] S. Aign and K. Fazel. Temporal & spatial error concealment techniques for hierarchical MPEG-2 video codec. In *Proceedings IEEE International Conference on Communications ICC'95*, Seattle, USA, vol. III, June 1995, pp. 1778-1783.

[4] S. Aign. Adaptive temporal & spatial error concealment measures in MPEG-2 video decoder with enhanced error detection. In Biglieri, E. and Luise, M. (Eds.). *Proceedings 7th Thyrrhenian Workshop on Digital Communications, Signal Processing in Telecommunications*, Springer-Verlag, Viareggio, Italy, September 1995, pp. 112-123.

[5] S. Aign. Error concealment, early re-synchronization, and iterative decoding for MPEG-2. In *Proceedings IEEE International Conference on Communications ICC'97*, Montréal, Canada, vol. III, June 1997, pp. 1654-1658.

[6] R.E. Blahut. *Theory and Practice of Error Control Codes*, Addison-Wesley Publishing Company, Inc., 1984.

[7] C.Le Buhan. Software embedded data retrieval and error concealment scheme for MPEG-2 video sequences. In *Proceedings Conference on Electronic Imaging, Digital Video Compression: Algorithm and Technologies SPIE'96*, San Jose, California, vol. 2668, February 1996, pp. 384-391.

[8] E.R. Davies. *Machine Vision*, Academic Press Inc., San Diego, 1990.

[9] "Digital broadcasting systems for television, sound and data services; Framing structure, channel coding and modulation for 11/12 GHz satellite services", ETS 300 421, 1994.

[10] "Digital broadcasting systems for television, sound and data services; Framing structure, channel coding and modulation for terrestrial systems", prETS 300 744, 1995.

[11] "Digital broadcasting systems for television, sound and data services; Framing structure, channel coding and modulation for cable systems", ETS 300 429, 1994.

[12] J. Feng, K.-T. Lo, H. Mehrpour, and A.E. Karbowiak. Loss recovery techniques for transmission of MPEG video over ATM networks. In *Proceedings IEEE International Conference on Communications ICC'96*, Dallas, USA, vol. III, June 1996, pp. 1406-1410.

[13] C.L. Fernández, A. Basso, and J.P. Hubaux. Error concealment and early resynchronization techniques for MPEG-2 video streams damaged by transmission over ATM networks. In *Proceedings SPIE'96*, San Jose, California, vol. 2668, February 1996, pp. 372-383.

[14] M. Ghanbari and V. Seferidis. Cell-loss concealment in ATM video codecs. In *IEEE Transactions on Circuits and Systems for Video Technology*, vol. 3, no. 3, June 1993, pp. 238-247.

[15] J. Hagenauer and P. Höher. A Viterbi algorithm with soft-decision outputs and its applications. In *IEEE Global Telecommunications Conference GLOBECOM'89*, Dallas, Texas, November 1989, pp. 47.1.1-47.1.7.

[16] P. Haskell and D. Messerschmitt. Resynchronization of motion compensated video affected by ATM cell loss. In *Proceedings IEEE International Conference on Acoustics, Speech, and Signal Processing ICASSP'92*, vol. 3, 1992, pp. 545-548.

[17] S.S. Hemami and H.Y. Meng. Spatial and temporal video reconstruction for non-layered transmission. In *Proceedings Fifth International Workshop on Packet Video Visicom'93*, March 1993.

[18] K.-H. Jung and C.W. Lee. Projection-based error resilience technique for digital HDTV. In *Proceedings HDTV-Workshop'94*, Turin, Italy, October 1994.

[19] W. Kwok and H. Sun. Multi-directional interpolation for spatial error concealment. In *IEEE Transactions on Consumer Electronics*, vol. 39, no. 3, August 1993, pp. 455-460.

[20] S. Lee, J.S. Youn, S.H. Jang, and S.H. Jang. Transmission error detection, resynchronization, and error concealment for MPEG video decoder. In *Proceedings SPIE'93*, vol. 2094, 1993, pp. 195-204.

[21] W. Luo and M.El Zarki. Analysis of error concealment schemes for MPEG-2 video transmission over ATM based networks. In *Proceedings SPIE'95*, vol. 2501, 1995, pp. 1358-1368.

[22] W.-M. Lam, A.R. Reibman, and B. Liu. Recovery of lost or erroneously received motion vectors. In *Proceedings IEEE International Conference on Acoustics, Speech, and Signal Processing ICASSP'93*, Minnesota, USA, vol. V, April 1993, pp. 417-420.

[23] X. Lee, Y.-Q. Zhang, and A. Leon-Garcia. Information loss recovery for block-based image coding techniques – a fuzzy logic approach. In *Proceedings SPIE'93*, vol. 2094, 1993, pp. 529-540.

[24] *Generic Coding of Moving Pictures and Associated Audio Information - Systems*, MPEG-2 Standard ISO/IEC 13818-1, November 1994.

[25] *Generic Coding of Moving Pictures and Associated Audio Information - Video*, MPEG-2 Standard ISO/IEC 13818-2, November 1994.

[26] A. Narula and J.S. Lim. Error concealment for an all-digital high-definition television system. In *Proceedings SPIE'93*, vol. 2094, 1993, pp. 304-315.

[27] B. Ramamurthi and A. Gersho. Nonlinear space-variant postprocessing of block coded images. In *IEEE Transactions on Acoustics, Speech, and Signal Processing*, vol. ASSP-34, no. 5, October 1986, pp. 1258-1268.

[28] H.R. Rabiee, H. Radha, and R.L. Kashyap. Error concealment of still image and video streams with multi-directional recursive nonlinear filters. In *Proceedings IEEE International Conference on Image Processing ICIP'96*, Lausanne, Switzerland, vol. 2, September 1996, pp. 37-40.

[29] H. Sun, K. Challapali, and J. Zdepski. Error concealment in digital simulcast AD-HDTV decoder. In *IEEE Transactions on Consumer Electronics*, vol. 38, no. 3, August 1992, pp. 108-118.

[30] P. Salama, N.B. Shroff, E.J. Coyle, and E.J. Delp. Error concealment techniques for encoded video streams. In *Proceedings IEEE International Conference on Image Processing ICIP'95*, Washington DC, October 1995, pp. 9-12.

[31] P. Salama, N.B. Shroff, and E.J. Delp. A Bayesian approach to error concealment in encoded video streams. In *Proceedings IEEE International Conference on Image Processing ICIP'96*, Lausanne, Switzerland, vol. 2, September 1996, pp. 49-52.

[32] H. Sun and W. Kwok. Error concealment with directional filtering for block transform coding. In *Proceedings IEEE Global Telecommunications Conference GLOBECOM'93*, Houston, Texas, December 1993, pp. 1304-1308.

[33] H. Sun and W. Kwok. Concealment of damaged block transform coded images using projections onto convex sets. In *IEEE Transactions on Image Processing*, vol. 4, no. 4, April 1995, pp. 470-477.

[34] H. Sun and J. Zdepski. Adaptive error concealment algorithm for MPEG compressed video. In *Proceedings Visual Communications and Image Processing SPIE'92*, vol. 1818, 1992, pp. 814-814.

[35] H. Sun, M. Uz, J. Zdepski, and R. Saint Girons. A proposal for increased error resilience. ISO-IEC/JTC1/SC29/WG11, MPEG 92/532, Sept. 1992.

[36] M. Wada. Selective recovery of video packet loss using error concealment. In *IEEE Journal on Selected Areas in Communications*, vol. 7, June 1989, pp. 807-814.

[37] Y. Wang and Q.-F. Zhu. Signal loss recovery in DCT-based image and video codecs. In *Proceedings SPIE Visual Communications and Image Processing SPIE'91*, Boston, USA, vol. 1605, Nov. 1991, pp. 667-678.

[38] Y. Wang, Q.-F. Zhu, and L. Shaw. Maximally smooth image recovery in transform coding. In *Transactions on Communications*, vol. 41, no. 10, October 1993, pp. 1544-1551.

[39] W. Zeng and B. Liu. Geometric-structure-based directional filtering for error concealment in image/video transmission. In *Proceedings SPIE Wireless Data Transmission, Photonics East'95*, vol. 2601, October 1995, pp. 145-156.

[40] W. Zhu and Y. Wang. The use of second order derivative based smoothness measure for error concealment in transform based codecs. In *Proceedings SPIE'95*, vol. 2501, 1992, pp. 1205-1214.

[41] Q.-F. Zhu, Y. Wang, and L. Shaw. Image reconstruction for hybrid video coding systems. In *Proceedings IEEE Data Compression Conference*, March 1992, pp. 229-238.

[42] Q.-F. Zhu, Y. Wang, and L. Shaw. Coding and cell-loss recovery in DCT-based packet video. In *IEEE Transactions on Circuits and Systems for Video Technology*, vol. 3, no. 3, June 1993, pp. 248-258.

9 COMBINED SOURCE-CHANNEL VIDEO CODING

King N. Ngan and C. W. Yap

Visual Communications Research Group
Department of Electrical and Electronic Engineering
The University of Western Australia
Crawley, W.A. 6907, Australia
{king,cwyap}@ee.uwa.edu.au

Abstract: In this chapter, we have discussed and outlined a novel approach in the design of an error resilient combined source-channel coding scheme for the transmission of video over highly error prone mobile channels. The chapter begins with an introduction to combined source-channel coding followed by a short overview of work performed in this area in the past. A scheme based upon the H.263 standard and modifications performed to it to facilitate the application of UEP are described. Additional sections cover the EREC algorithm used to preserve synchonisation, and RCPC codes used for UEP. We have demonstrated our approach over various channel models ranging from Gaussian to Rayleigh with the best performance gain obtained using a soft decision channel decoder and the EREC algorithm. In addition, we have also shown that the H.263 decoder requires only simple modifications to provide a measure of error concealment such that certain visually annoying artefacts may be removed.

Acknowledgments

This work is supported in part by the Australian Cooperative Research Centre for Broadband Telecommunications and Networking.

9.1 INTRODUCTION

The transmission of high bandwidth video sequences over mobile communication channels presents several challenging problems that remain to be resolved. Firstly, there is the problem of the source bandwidth. This is easily reduced by source coding it with one of the many algorithms such as sub-band coding, H.261 [1], MPEG2 [2], and H.263 [3], which are in turn based upon mathematical transforms such as the Discrete Cosine Transform (DCT). These algorithms generally code the image sequence spatially and temporally. Spatial coding is performed using the DCT, while temporal redundancy is reduced using motion compensation and prediction (interframe coding). The source coded data are then Huffman coded, layered and packed with headers before being output.

Secondly, the mobile channel impedes the correct transmission and reception of highly compressed video sequences by introducing burst errors caused by multipath fading of the transmitted signal. The multipath fading occurs due to the reflection of the transmitted signal off building walls and other animate or inanimate objects, and the movement of the mobile receiver through such an environment. In the worst cases, Rayleigh fading is obtained whereby there is no direct line of sight between the transmitter and the receiver. At other times when there is a direct path present in addition to the reflected paths, the fading present is known as Rician fading.

To reduce the number of errors introduced by the channel, channel coding is employed. Generally, the channel coding used is based upon the average or worst case received bit error rate required for a particular channel or environment. Thus, a fixed code with a certain rate is applied uniformly over the source coded data before transmission. This then provides a guaranteed level of error control. However, this is relatively inefficient because generally, different source coded bits require different levels of error protection. Therefore, the source coder has to be modified to provide information about the significance of the source bits which can then be utilised to adjust the level of error protection applied by the channel coder. In this chapter, we look at using Rate Compatible Punctured Convolutional (RCPC) codes to provide Unequal Error Protection (UEP) in the mobile environment.

Since a reduction of channel errors does not necessarily preclude synchronisation errors from occurring during the source decoding stage, an error resilient synchronisation technique has to be applied to prevent catastrophic errors from causing a loss of synchronisation. One such technique that does not increase the bit rate considerably is known as the Error Resilient Entropy Coder (EREC) [4]. It re-arranges the variable length Huffman codewords into fixed length blocks before channel coding is applied. This results in the preservation of synchronisation at the receiver end. In the event that the channel bit error rate (BER) increases, the use of the EREC algorithm results in graceful deterioration instead of catastrophic failure. In addition, this chapter presents several error concealment techniques implemented in the source decoder to remove

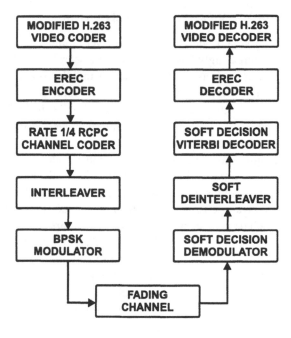

Figure 9.1   Block diagram of proposed system.

certain residual errors that remain after channel decoding. A block diagram of the system described in this chapter is shown in Fig. 9.1.

The organisation of the rest of this chapter is as follows: Section 9.2 is a short overview of research leading to the area of combined source-channel coding. Section 9.3 briefly introduces the H.263 codec and describes the modifications performed on the source coder to produce source significant information (SSI) required by the channel coder. Section 9.4 covers the EREC algorithm as applied in this chapter and Section 9.5 describes the RCPC codes used in channel coding and how the source coded bit stream is segmented into different significance classes. Section 9.6 covers the error concealment strategy and Section 9.7 briefly covers the channel model and channel simulation while Section 9.8 presents the results of simulations based upon a combined source-channel coding scheme. In the course of this chapter, we will combine various subsystems to create a scheme for the transmission of mobile video using combined source-channel coding.

9.2 OVERVIEW OF PREVIOUS RESEARCH WORK

The coding and transmission of video information over mobile communication channels has emerged as a increasingly interesting area of research over the past ten years. This is because of factors such as the acceptance of mobile communications by the public at large, and an improvement in the systems and technology used. This makes the long term goal of transmitting real time video over mobile channels more feasible.

Prior to research on joint source-channel coding, they were each treated as separate processes where information transmission was concerned. In 1979, Modestino and Daut used a two-dimensional (2-D) differential pulse coded modulation (DPCM) technique for source coding and various rates of convolutional codes for channel coding [5]. Their main aim of joint source-channel coding is to protect some of the coded bits that are more sensitive to channel errors than other bits. This led on to bit error sensitivity studies which have been carried out for both compressed speech and video based upon the type of source coding used [6, 7, 8]. These bit error sensitivity studies involve the insertion of bit errors into consecutive bits within the source coded data, which are then source decoded, and the resultant image compared with the original using some form of objective measure like peak signal to noise ratio (PSNR). This will then enable the system designer to tailor the level of error protection applied to individual bits, or sets of bits, to obtain an unequal error protection system.

Work on the transmission of still image data over low bandwidth wireless links can be found in [9, 10]. The first paper concentrates mainly on the transmission of facsimile images over an analogue FM channel with FEC and ARQ, while the second paper uses a 1.5 GHz carrier based digital mobile channel with FEC and time diversity. A performance study on still image data over mobile channels was carried out by Brewster [11] to show the various combinations of using FEC and ARQ over mobile channels with different burst error length distributions. The move from still image to video sequences took off with the introduction of the ITU-T H.261 standard [1] for video transmission at $p \times 64$ Kbps (from 64 Kbps to 2 Mbps). This allowed the development of systems transmitting video over wired channels, such as Integrated Services Digital Network (ISDN) channels. The H.261 is a spatiotemporal scalable coder in the sense that better quality and/or higher frame rate can be obtained when more bandwidth is available. Following the development of H.261, there appeared a possibility of using it to transmit compressed video over Digital European Cordless Telecommunications (DECT) channels that have a maximum capacity of 2 Mbps [12]. This allowed a certain degree of bandwidth overlap between the H.261 coder and the DECT channel. Transmission using other mobile channels, such as the Japanese Personal Handyphone System (PHS) channels, have also been investigated [13].

In the field of information theory, Clark and Cain applied the concept of puncturing, the systematic removal of certain channel coded bits, to convolutional codes which resulted in a set of codes whose rates can be modified to provide differing levels of error protection [14]. These codes, known as RCPC codes, are described further in section 9.5.1. Hagenauer later discovered a set of RCPC codes that has as good a level of error protection as convolutional codes with fixed rates [15]. Evidently, these codes can applied naturally to joint source-channel coding applications. Since a lot of compressed video data have varying bit error sensitivities, Pelz implemented a $p \times 8$ Kbps ($p = 1 \ldots 4$) over DECT channels using RCPC codes [16]. Other types of FEC codes used for joint source-channel coding are linear unequal protection codes, as shown by Fazel and L'Huillier [8].

Applications for joint source-channel coding of video range from the multiresolution broadcast of high definition television to adaptive low-rate wireless videophone schemes [17, 18, 19, 20].

9.3 SOURCE CODING

This section starts by describing the operation of the H.263 codec briefly, followed by a description of the ITU-T H.263 bitstream syntax. It concludes by detailing the modifications to the H.263 codec that enable our combined source-channel coding scheme to work.

9.3.1 H.263 video coding algorithm

The ITU-T Experts Group for Very Low Bit-Rate Video Telephony (LBC) has produced a set of near-term recommendations for video telephony over PSTN (Public Switched Telephony Network) aimed at transmission at less than 64 Kbps. It is known as the ITU-T Draft Recommendation H.263 and is finalised in 1996 [3]. The H.263 algorithm is an extension of the well-known H.261 video coding algorithm for videoconferencing at $p \times 64$ Kbps [1]. It is based on the same motion compensated hybrid DPCM/DCT coder with considerable improvements to achieve bit rates less than 64 Kbps, as shown in Fig. 9.2.

9.3.2 Video formats

The input video format in the source coder is a common intermediate format (CIF) having a spatial resolution of 288 lines by 352 pixels per line, to cater for both the 625- and 525-line television standards. The CIF format can be downsampled to QCIF format having a quarter of the spatial resolution of the CIF format. Other video formats acceptable are the sub-QCIF (96 lines by 128 pixels per line), 4CIF (576 lines by 704 pixels per line), and 16CIF (1152 lines by 1408 pixels per line). The encoder must be able to operate with either QCIF or sub-QCIF formats, others being optional. But all decoders must be able to decode QCIF and sub-QCIF pictures. Colour formats and position of luminance and chrominance samples are the same as H.261.

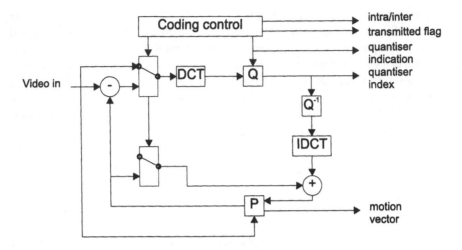

Figure 9.2　The H.263 source coder.

Each picture is divided into groups of blocks (GOBs). A GOB comprises $k \times 16$ lines, where k depends on the picture format. The number of GOBs per picture is 6 for sub-QCIF, 9 for QCIF, and 18 for CIF, 4CIF and 16CIF. The GOB numbering is done by the use of vertical scan of the GOBs, starting with the upper GOB and ending with the lower GOB.

Each GOB is divided into macroblocks (MBs). A macroblock comprises a block of 16×16 pixels of luminance (Y) and two spatially corresponding blocks of 8×8 pixels of chrominance (C_B and C_R). Therefore, it consists of four Y blocks, a C_B and a C_R block in the order as shown in Fig. 9.3.

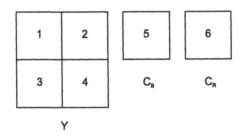

Figure 9.3　Arrangement of blocks in a macroblock.

9.3.3 Source coding algorithm

The coding of the source images is block-based. Two operation modes namely intra and inter modes, are used to compress the image data to the desirable bit rates for transmission, as can been seen in Fig. 9.2. Intra mode reduces the spatial redundancy within the picture; whilst the inter mode is used to reduce the temporal redundancy between pictures. The first picture (I-picture) is always intra-coded by transforming the pixel blocks using DCT (discrete cosine transform). Subsequent pictures (P-pictures) are inter-coded by transforming the prediction differences between the current picture and its motion-compensated (MC) prediction from the previous picture. The transform coefficients are then quantised and variable length coded (VLC) using a pre-defined code table.

The decision to code a macroblock intra or inter, and with MC or without MC, are determined by specific algorithms not subject to standardisation.

9.3.4 Video bitstream syntax

The H.263 bitstream syntax is a 4-layered structure: picture layer, group-of-blocks layer, macroblock layer, and block layer as shown in Fig. 9.4.

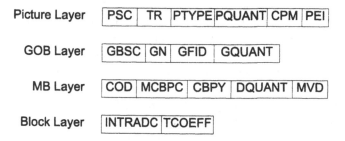

Figure 9.4   The H.263 bitstream syntax.

Picture layer. A picture layer consists of a picture header followed by data for a GOB layer. At the end of a sequence, a GOB layer is followed by an EOS (end of sequence) code and a number of PSTUF (stuffing) bits. The picture start code (PSC) is a 22-bit codeword. The 8-bit temporal reference (TR) is formed by incrementing its value in the previously transmitted picture header by one plus the number of non-transmitted pictures since the previously transmitted one and it addresses the P-pictures. The 13-bit picture type information (PTYPE) contains information about the picture such as split screen indicator, document camera indicator, freeze picture release, source format, picture coding type, and the use of optional special coding mode. There are four special coding modes:

- Unrestricted motion vector mode;

- Syntax-based arithmetic coding mode;

- Advanced prediction mode;

- PB-frames mode.

A fixed-length codeword (PQUANT) of 5 bits indicates the quantiser to be used for the picture unless updated by any subsequent GQUANT or DQUANT. The PQUANT value represents half the step sizes of the quantiser to be used. The 1-bit codeword (CPM) signals the use of the optional continuous presence multipoint mode. The extra insertion information (PEI) bit is set to 1 when there is a following optional data field contained in the spare information (PSPARE) bits.

Group of blocks layer. The group of block layer consists of a GOB header followed by data for the macroblocks. For the first GOB in each picture (with number 0), no GOB header shall be transmitted. The group of block start code (GBSC) is a 17-bit codeword and may be byte-aligned. The number of GOBs is indicated by a 5-bit group number (GN) codeword. GOB frame ID (GFID) is a 2-bit codeword having the same value in every GOB header of a given picture. The quantiser to be used for the remaining part of the picture can be changed by setting a 5-bit codeword GQUANT until updated by any subsequent GQUANT or DQUANT. As in the case of PQUANT, GQUANT value represents half the step sizes of the quantiser to be used.

It is left as an option that every GOB except the first, begins with a GOB header containing the following: GBSC, GN, GFID, and GQUANT. The first GOB has a GBSC and GN which is the mandatory PSC. If we are able to remove the GOB headers before transmission and use the EREC decoder to re-introduce them at the receiving end, we would obtain a savings in the number of bits transmitted and also regain synchronisation for the H.263 decoder.

Macroblock layer. Data for each macroblock consists of a macroblock header followed by data for the blocks. If the coded macroblock indication (COD) bit is set to '0', the macroblock is coded. It it is '1', no further information is transmitted for this macroblock. The MCBPC is a variable length codeword giving information about the macroblock type and coded block pattern for chrominance and is always included in coded macroblocks. The variable codewords are pre-defined in tables for both I-pictures and P-pictures. An extra codeword is used for bit stuffing. Similarly, CBPY is a pre-defined variable codeword for luminance. Again, different tables are used for I-pictures and P-pictures.

Quantiser information can be updated by a 2-bit DQUANT codeword which defines a change in quantiser value. Motion vector data (MVD) is included for all inter macroblocks and is variable length coded.

Block layer. INTRADC is a 8-bit codeword used to code DC coefficient of the intra blocks. Other transform coefficients are coded using variable length codewords (TCOEF) which represents an EVENT. An EVENT is a combination of last non-zero coefficient indication (LAST), the number of successive zeroes preceding the coded coefficient (RUN), and the non-zero value of the coded coefficient (LEVEL) along a zig-zag scan path as shown in Fig. 9.5.

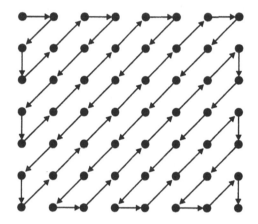

Figure 9.5   Zig-zag scan path of transform coefficients.

9.3.5 Modifications to H.263 coder

The first modification to the H.263 coder is to enable it to output a secondary bit stream as input into the channel coder. This bit stream contains source significance information (SSI) which controls the puncturing of the channel coded data. It is not transmitted across the channel. Further details about the SSI data are provided in Section 9.5.2. The second modification to the H.263 coder is to allow it to supply the EREC coder with the number of bits contained in each MB. This will be used by the EREC coder to reorganise the variable length blocks of data into fixed length blocks as explained in the next section. In addition, since the EREC algorithm takes care of MB synchronisation, the GOB headers are not transmitted.

9.4 ERROR-RESILIENT ENTROPY CODING

The EREC algorithm [4] is an entropy based error resilient coder. Since the output of the H.263 coder is made up of variable length codewords, serial transmission of such data through an error prone channel would cause errors to propagate through the bit stream. This will cause the source decoder to lose synchronisation and thus may result in decoded video frames being spatially corrupted. In order to maintain synchronisation, the most obvious solution is to insert synchronisation codewords that will increase the bit length. Because the longest H.263 codeword is 13 bits in length, the synchronisation word used would have to exceed this length and must be robust enough to be detected in the presence of bit errors. This would result in an increase in the bit rate which would in turn reduce the frame rate in a bandwidth constrained channel. The EREC algorithm resolves this by reordering the variable length codewords into

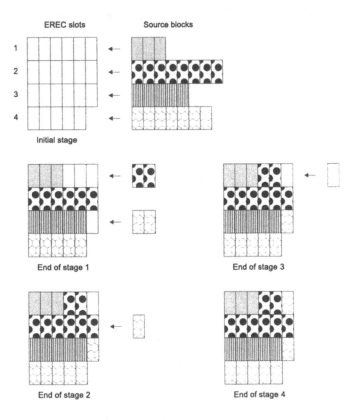

Figure 9.6   EREC bit re-ordering.

fixed length slots of data which are channel coded prior to transmission. Fig. 9.6 shows a graphical example of how the EREC reordering works. We shall explain the EREC algorithm using an imaginary source coder that outputs N_b variable length blocks of data with each block being of length b_i bits. In our implementation, the EREC coder outputs N_s slots of data each of length s_j bits. The total number of bits output by the EREC coder is $T_s = \sum_{j=1}^{N_s} s_j$ bits, and the total number of bits output by our source coder is $T_b = \sum_{i=1}^{N_b} b_i$ bits. In our implementation, we have chosen $N_s = N_b$ and $T_s = T_b$ which results in

$$s_j = \begin{cases} \tilde{s} + 1 & 1 \leq j \leq \text{remainder}\left(\frac{T_s}{N_s}\right) \\ \tilde{s} & \text{remainder}\left(\frac{T_s}{N_s}\right) < j \leq N_s \end{cases}, \tag{9.1}$$

where $\tilde{s}$ denotes the integer part of T_s/N_s. Therefore, if our imaginary source coder has $T_b = 23$, and $N_b = 4$ with the following block lengths $b_i = \{3, 8, 5, 7\}$, from Eq. (9.1), we obtain $\tilde{s} = 5$ and

$$s_j = \begin{cases} 6 & 1 \leq j \leq 3 \\ 5 & 3 < j \leq 4 \end{cases}.$$

With reference to Fig. 9.6, re-ordering of the bits from the variable length blocks into the fixed length slots occurs with the following steps:

The EREC bit re-ordering algorithm

```
{
    assign slot sⱼ to block bᵢ
    for each block bᵢ
    {
        if block bᵢ has bits left to transfer
        {
            if slot sⱼ is not full
                transfer as many bits as possible from bᵢ to sⱼ
            else
                assign slot sⱼ₌ⱼ₊φₙ  (modNₛ) to block bᵢ
        }
    }
}
```

In the above algorithm, ϕ_n is a predefined sequence. In our implementation, ϕ_n is a pseudo-random sequence that ranges from 1 to N_s. It has been found that a pseudo-random slot order is preferable and results in quicker decoding due to its uncorrelated nature [4]. From Fig. 9.6, the progression of the EREC can be followed till its completion. In our implementation, N_s and N_b are fixed at the number of MBs

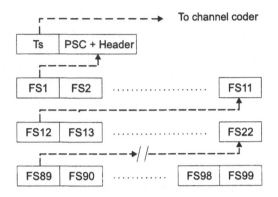

Figure 9.7   Output of EREC coder.

in a QCIF frame; which is 99. The only information required at the EREC decoder is T_s, since N_s and ϕ_n are already known information.

The EREC decoder works in reverse to decode the variable length blocks of data from the fixed length EREC slots. The only complexity in the implementation of the decoder is that it must be able to recognise when each MB has been fully decoded. This entails the inclusion of part of the H.263 decoder within the EREC decoding algorithm. In addition, our implementation of the EREC decoder inserts a GOB header after every 11 MBs decoded. This results in a saving of 176 bits per frame and this enables us to transmit a highly protected value of T_s across the channel.

The EREC coder outputs in the following order, T_s, H.263 header, and the fixed length slots (FS1 to FS99), as shown in Fig. 9.7. T_s is marked with the highest significance class so that the channel coder will code it fully. We may also utilise a Reed-Solomon code as an inner code, in a concatenated coding scheme, to protect T_s before passing it on to the RCPC coder.

9.5 CHANNEL CODING

Implementation of channel coding in our system uses punctured convolutional codes which are described, following a short introduction to convolutional codes, in Section 9.5.1. These codes are then applied to the output of the EREC coder using the method described in Section 9.5.2.

9.5.1 RCPC codes

All channel coders take source data either in the form of bits or words (groups of bits) and convert them into codewords which have additional redundancy to protect the source data from channel errors. Block based channel coders take the source bit stream and segment them into fix length blocks, which are then mapped onto their corresponding codewords. A convolutional coder converts the input datastream into one continuous codeword. Essentially, the convolutional coder takes the input data and discretely convolves it with a generator matrix where the current output bit is formed by a modulo-2 sum of previously delayed input bits. This is usually performed by passing the input bits through a linear finite state shift register with several taps combined in modulo-2 to obtain the output bits (see Fig. 9.8). It is possible to segment the input bitstream into fix length blocks provided the shift register of length m is "flushed" by the addition of m tail bits consisting of binary zeros at the end of each block. These tail bits are required to clear the contents of the shift register to zero before the input of the next block of bits.

The code rate of a convolutional encoder is defined as the ratio k/n, where k and n are the number of inputs (shift registers) and outputs (modulo-2 adders), respectively. The convolutional encoder we used outputs 4 bits for every input bit, thus making it a rate $1/4$ coder. One of the properties of convolutional codes is that low rate codes can be punctured to produce codes of higher rates [14]. This makes it very attractive for applications where variable code rates are required.

Puncturing is defined as the act of deleting one or more bits from the output of the encoder's bit stream. Because the code rate is adjustable, these codes are known as RCPC codes [15]. In addition to the variable rate, simplified maximum likelihood decoding can be obtained with certain families of RCPC codes. These non-catastrophic RCPC codes also offer a comparable level of error protection to fixed convolutional codes at similar rates. RCPC codes are thus suited for combined source-channel coding of information where the error protection levels of the source bits are modulated by the significance of the bits themselves.

Table 9.1   Generator polynomials and puncturing matrices for RCPC codes.

Generator polynomials	Puncturing matrix		
	Rate 1/3	Rate 1/2	Rate 2/3
$\mathbf{g}^{(0)} = 1 + D^3 + D^4$	11111111	11111111	11111111
$\mathbf{g}^{(1)} = 1 + D + D^2 + D^4$	11111111	11111111	10101010
$\mathbf{g}^{(2)} = 1 + D^2 + D^3 + D^4$	11111111	00000000	00000000
$\mathbf{g}^{(3)} = 1 + D + D^3 + D^4$	00000000	00000000	00000000

To illustrate coding with RCPC codes, we will use our system as an example. In our implementation, a mother code of rate $1/4$ is used to protect the source bits at the lowest rate. The output bits are then punctured based upon the SSI provided by the H.263 source coder. These punctured bits are removed and not transmitted across the channel. The generator polynomials used to generate the codes and the puncturing matrices are shown in Table 9.1 with the corresponding implementation shown in Fig. 9.8. Each '0' in the puncturing matrix represents a bit that is punctured.

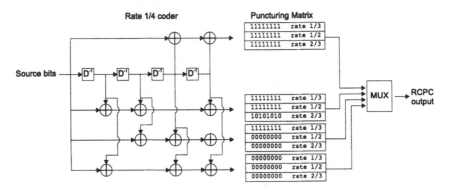

Figure 9.8   The RCPC encoder.

From the puncturing matrices, it can be observed that the puncturing occurs with a period of 8 bits. After transmission through the channel, the noise degraded signal is passed through the demodulator which then can either make a hard or a soft decision on the received value. A hard decision results from a quantisation of the k-th received symbol, r_k, to binary levels as shown,

$$d_k = \begin{cases} 1 & r_k > 0 \\ 0 & r_k \leq 0 \end{cases},$$

(9.2)

while a soft decision makes use of channel information contained in the unquantised received signal to aid in the decoding process. Practically however, in soft decision decoding, r_k is quantised to n levels,

$$d_k = \begin{cases} d_1 & r_k \leq l_1 \\ d_2 & l_1 < r_k \leq l_2 \\ \vdots & \vdots \\ d_n & r_k > l_{n-1} \end{cases},$$

(9.3)

where d_k is the k-th output of the decision circuit, with quantisation intervals from l_1 to l_{n-1}.

The decoding algorithm used is the well known maximum likelihood Viterbi algorithm [21] with 3 bit quantisation used to improve the decoder's performance over hard decisions. It has been shown that for additive white Gaussian noise channels, a Viterbi decoder with a 3 bit quantiser approaches the performance of one using unquantised outputs to within 1 dB [22]. Unlike Reed-Solomon codes, RCPC codes are not good at handling errors of a bursty nature. Therefore after puncturing, interleaving is performed to randomise the burst errors. Correspondingly, a de-interleaver is used at the decoder. The level of interleaving is dependant on the average duration of fade produced by the channel. For a Rayleigh channel, the average duration of fade is defined as a function of the received signal envelope and is given here by

$$\overline{\tau} = \frac{\exp\left(\rho^2\right) - 1}{\rho f_d \sqrt{2\pi}} \, , \tag{9.4}$$

where ρ is the level of fade, f_d is the Doppler frequency as defined in Section 9.7, and $\overline{\tau}$ is in seconds.

9.5.2 Unequal error protection

This section describes the UEP [23] used to segment the source coded H.263 data into 3 classes of significance [24]. The source bits are channel coded using the rate 1/4 mother code and later punctured to obtain the desired rate. The puncturing will depend on the classification of the bits. We have divided the H.263 bit stream into three layers as shown in Fig. 9.9, with Class 2 being the most important and Class 0 the least important. It can be seen that the most important bits are those that cause the PSNR to

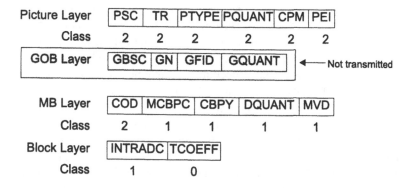

Figure 9.9   Classification of the H.263 bit stream.

degrade the most, such as the header and synchronisation bits which tend to garble the video sequence completely. The least important bits are those that will cause a lesser degradation in the received PSNR. More important class 2 bits are transmitted at the lowest rate, with less puncturing, while least important class 0 bits are transmitted at the highest rate, due to increased puncturing.

9.6 ERROR CONCEALMENT

Residual errors remaining in the channel decoded H.263 bitstream can result in artefacts being introduced into the resultant video sequence. Some simple error concealment procedures have been implemented in the basic H.263 decoder to remove such artifacts. If error concealment is not implemented, it is very likely that the H.263 decoder would fail.

The first simple error concealment procedure is implemented on two layers - the picture layer and the GOB layer. It depends solely on the detection of an error usually consisting of one of the following types:

- VLC not found in VLC table,

- PSC not found or in error.

These errors can occur due to a loss of synchronisation within the H.263 bit stream, or modification of the Huffman codewords by residual errors. A change in a single bit will result in the source decoder erroneously decoding the codeword. This can cause the decoding of the current GOB to end prematurely or to overrun into the next GOB. When this occurs, the modified H.263 decoder will search the incoming bit stream until a valid GBSC and GN are found. The GN will then be used as a guide for the decoder to replace the current GOB slice(s) in error with the GOB(s) from the previously decoded frame. If the concealment algorithm detects the PSC (which is actually a GOB header with a GN of zero) from the next frame, then the entire frame is replaced by the previously decoded one. Because the GOB headers are inserted by the EREC decoder at the receiver, this error concealment method works very well.

The second error concealment procedure is implemented to remove the out of range transform coefficients, which can result in high frequency artefacts producing a checkerboard pattern after the inverse DCT operation. These coefficients should lie in the range of -256 to +255. If any of these coefficients are out of range, they are set to zero. Fig. 9.10 shows one frame of a sequence decoded with and without the coefficient error concealment implemented. In Fig. 9.10, we observe in the magnified nose region of the *Alexis* sequence, the checkerboard pattern that appears when these coefficients are decoded out of range.

9.7 MOBILE CHANNEL MODEL

This section describes the mobile channel model employed to test the combined source-channel coder. In the mobile channel environment, the multipath propagation from

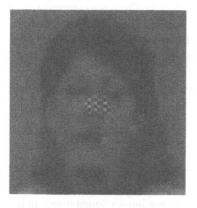

Figure 9.10   Magnified single frame of *Alexis* sequence without coefficient error concealment (*left*), and with error concealment (*right*).

the source to the receiver causes the received signal envelope to fluctuate in a Rician manner, with possibly one or more direct line of sight paths and several reflected paths [25, 26] as shown in Fig. 9.11. When there is no line of sight (ray a) from the transmitter to the receiver, the fading obtained is Rayleigh fading, else it is Rician fading.

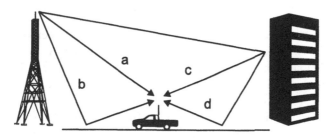

Figure 9.11   Multipath effect.

The probability density function exhibited by the received signal is given by

$$p(r) = \begin{cases} \frac{r}{\sigma^2} \exp\left(-\frac{r^2+A^2}{2\sigma^2}\right) I_0\left(\frac{Ar}{\sigma^2}\right) & 0 \leq r \leq \infty \\ 0 & r < 0 \end{cases}, \qquad (9.5)$$

where σ^2 is the time averaged power of the signal before detection, A is the peak amplitude of the dominant signal, and $I_0\left(\cdot\right)$ represents the modified Bessel function of the first kind, zero-th order. When $A \to 0$, we obtain the Rayleigh probability density function.

The channel simulator used to simulate a mobile unit travelling at a speed of v m/s using a carrier of f_c Hz is based upon the basic model first implemented by Smith [27]. This model, with some modifications, is shown in Fig. 9.12. Two baseband Gaussian noise generators are lowpass filtered with filters having the following spectral response:

$$S(f) = \frac{1.5}{\pi f_d \sqrt{1 - \left(\frac{f - f_c}{f_d}\right)^2}} , \qquad (9.6)$$

where f_d is the Doppler frequency given by $f_d = v/\lambda$, λ is the wavelength of the carrier, and f is in the range of $f_c \pm f_d$. Eq. (9.6) is a representation of the fading spectrum of a vertical $\lambda/4$ antenna [28]. The filters are implemented using an inverse FFT operation on N Gaussian samples weighted by the power spectrum $S(f)$. The number of tones in the power spectrum corresponds to the product of the normalised fading bandwidth and N the number of points used in the inverse FFT and is given by $(f_d/f_s) \times N$, where f_s is the baseband symbol rate. The complex envelope of the fading process is given by

$$Z_k = I_k + jQ_k . \qquad (9.7)$$

The variance and mean of the process, σ^2 and A, respectively, are adjusted with the following constraints:

$$A = \sqrt{\frac{K}{K+1}} , \; \sigma^2 = \frac{1}{2(K+1)} , \; \text{and } K = \frac{A^2}{2\sigma^2} , \qquad (9.8)$$

where K is the Rician parameter defined as the ratio of the direct to diffuse signal power. When $K = 0$, Rayleigh fading is obtained, and when $K \to \infty$, we have a Gaussian channel. The above constraints will ensure that $E\{|Z_k|^2\} = 1$ which allows us to control the received energy per bit to noise ratio, or E_b/N_o, by varying the noise power. The BER, or p_e, of a Rayleigh channel is related to the type of modulation and the E_b/N_o by the following formula (shown for BPSK modulation)

$$p_e(\overline{\gamma}_b) = \frac{1}{2}\left(1 - \sqrt{\frac{\overline{\gamma}_b}{1 + \overline{\gamma}_b}}\right) , \qquad (9.9)$$

where

$$\overline{\gamma}_b = \frac{E_b}{N_o} E\{|Z_k|^2\} . \qquad (9.10)$$

The incoming baseband signal s_k is split into in-phase and quadrature-phase components which are then multiplied with the Rayleigh fading signal. They are then recombined and the noise parameter n_k is added. The output r_k is then presented to the receiver which either makes a hard or a soft decision. In our simulation, an 8-level soft decision demodulator is employed.

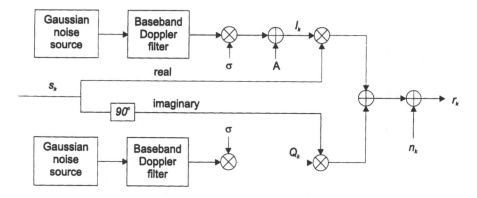

Figure 9.12   Channel simulator.

9.8   SIMULATION AND RESULTS

9.8.1   *Simulation*

Two QCIF sequences of 200 frames each are used in the simulation and the coding parameters for both sequences are shown in Table 9.2. Due to the high motion nature of the *Foreman* sequence, the target bit rate of the H.263 coder is increased to 30 Kbps while the low motion *Grandma* sequence is coded at the reduced bit rate of 9.6 Kbps. Each coded sequence is transmitted through Gaussian, Rician, and Rayleigh channels with BPSK modulation. We have simulated each sequence 50 times, with a mobile velocity of 6.4 km/h (walking speed), and for the Rician channel, with the Rician factor $K = 4$. At the mobile receiver, a 3-bit soft decision Viterbi decoder is used for all simulations. A comparison based upon the received PSNR Y, averaged over 50 runs, is made with the original sequences transmitted without any channel errors. Only PSNR Y is shown as it can be generally assumed that if the PSNR Y degrades, the PSNR of the C_B and C_R components will also be likely to degrade by a similar amount.

Results are also presented for the same sequences without the benefit of EREC coding to show the amount of improvement gained by using the EREC algorithm. All the sequences are decoded using the modified H.263 decoder with error concealment. During our simulations, we have assumed that the first frame is transmitted across intact as a reference for the H.263 decoder. This scheme does not take into account the method used to ensure that the first frame arrives at the decoder error-free. An ARQ or handshaking method can be used where, if any error at all is detected at the receiver end from any component, then the receiver can request for a retransmission with either higher transmitter power, better error protection, or a combination of both methods.

Table 9.2   H.263 simulation parameters.

Parameter	Grandma		Foreman	
Picture format	QCIF 176 × 144			
Number of frames	200			
Target bit rate (Kbps)	9.6		30	
Target frame rate (fps)	5.0			
Quantisation levels	16			
EREC algorithm used	No	Yes	No	Yes
GOB headers inserted	Yes	No	Yes	No
Transmitted size in bytes	44416	48448	119104	121920

9.8.2 Results

For all three channels with BPSK modulation, results for *Grandma* and *Foreman* sequences are given in Figs. 9.13 and 9.14. From the PSNR curves, it can be seen that the sequences using the the combined source-channel coding scheme together with the EREC perform better than without the EREC applied. This is more prominent for the Rician and Rayleigh channels than for the Gaussian channel as the length and frequency of burst errors increases in the former two channels. If the channels under consideration exhibit lower BERs than those shown here, then the RCPC puncturing matrix can be correspondingly adjusted to provide higher rate codes which would then allow a higher frame throughput.

Examining the PSNR graphs does not provide a subjective level for comparison. Therefore, included here are selected frames from the sequences simulated under these conditions. The original frames and those transmitted without channel errors are shown Fig. 9.15 and Fig. 9.16, respectively. Those transmitted via Gaussian, Rician and Rayleigh channels are shown in Figs. 9.17–9.22. From the images, we can compare the image degradation due to channel errors at the stated BER for those transmitted using our proposed method with and without the EREC algorithm.

The images also demonstrate the error concealment scheme in operation where a previous frame displayed in place of the current frame. This can be seen in Figs. 9.19, 9.20, and 9.22, where the frames displayed do not match the original source frames.

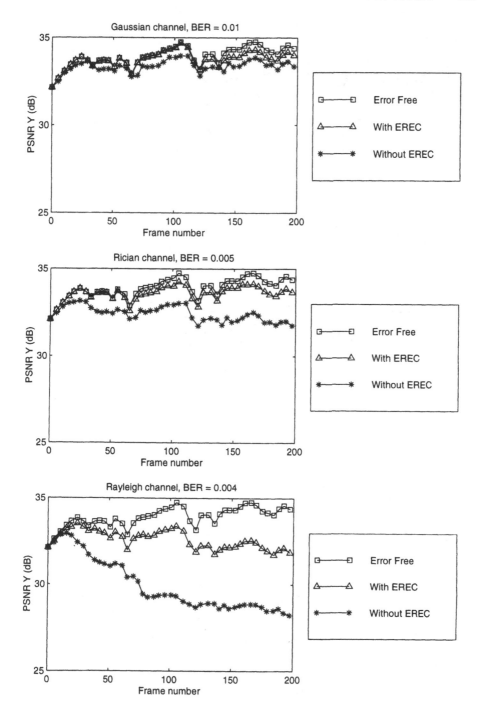

Figure 9.13   PSNR Y for *Grandma* sequence in a Gaussian channel with (*top*), Rician channel with $K = 4$ (*middle*), and Rayleigh channel with (*bottom*).

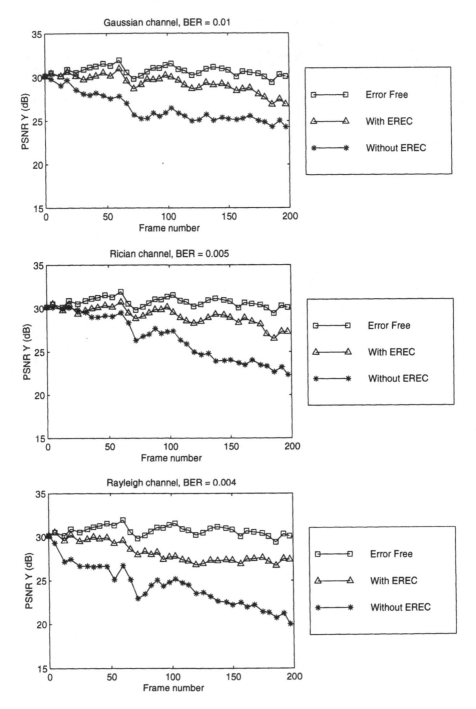

Figure 9.14   PSNR Y for *Foreman* sequence in a Gaussian channel with (*top*), Rician channel with $K = 4$ (*middle*), and Rayleigh channel (*bottom*).

Figure 9.15   Original frames - 131 of *Grandma* (*left*) and 83 of *Foreman* (*right*).

Figure 9.16   Transmitted frames without errors - 131 of *Grandma* (*left*) and 83 of *Foreman* (*right*).

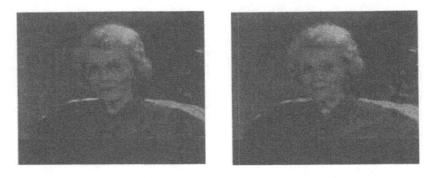

Figure 9.17 Gaussian channel *Grandma* frame 131, BER $= 0.01$, with EREC (*left*), without EREC (*right*).

Figure 9.18 Gaussian channel *Foreman* frame 83, BER $= 0.01$, with EREC (*left*), without EREC (*right*).

Figure 9.19 Rician channel *Grandma* frame 131, BER $= 0.005$, with EREC (*left*), without EREC (*right*).

Figure 9.20 Rician channel *Foreman* frame 83, BER $= 0.005$, with EREC (*left*), without EREC (*right*).

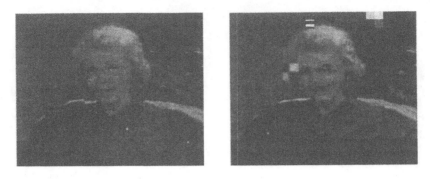

Figure 9.21 Rayleigh channel *Grandma* frame 131, BER $= 0.004$, with EREC (*left*), without EREC (*right*).

Figure 9.22 Rayleigh channel *Foreman* frame 83, BER $= 0.004$, with EREC (*left*), without EREC (*right*).

9.9 SUMMARY

In this chapter, we have discussed and outlined a novel approach in the design of an error resilient combined source-channel coding scheme for the transmission of video over highly error prone mobile channels. The chapter begins with an introduction to combined source-channel coding followed by a short overview of work performed in this area in the past. A scheme based upon the H.263 standard and modifications performed to it to facilitate the application of UEP are described. Additional sections cover the EREC algorithm used to preserve synchonisation, and RCPC codes used for UEP. We have demonstrated our approach over various channel models ranging from Gaussian to Rayleigh with the best performance gain obtained using a soft decision channel decoder and the EREC algorithm. In addition, we have also shown that the H.263 decoder requires only simple modifications to provide a measure of error concealment such that certain visually annoying artefacts may be removed.

Acknowledgments

This work is supported in part by the Australian Cooperative Research Centre for Broadband Telecommunications and Networking.

References

[1] ITU-T, *Video Codec for Audio Visual Services at p × 64 kbits/s*, March 1993.

[2] ISO/IEC/JTC1/SC29, *Generic Coding of Moving Pictures and Associated Audio Information: Video*, March 1995.

[3] ITU-T, *Video Coding for Low Bitrate Communication*, May 1996.

[4] D. W. Redmill and N. Kingsbury, "The EREC: An error resilient technique for coding variable-length blocks of data," *IEEE Transactions on Image Processing*, vol. 5, no. 4, pp. 565–574, April 1996.

[5] J. W. Modestino and D. G. Daut, "Combined source-channel coding of images," *IEEE Transactions on Communications*, vol. COM-27, no. 11, pp. 1644–1659, November 1979.

[6] R. V. Cox, J. Hagenauer, N. Seshadri, and C. E. Sundberg, "Combined source and channel coding using a sub-band coder," in *Proceedings of the 1988 IEEE International Conference on Communications*, 1988, pp. 1395–1399.

[7] H. Shi, P. Ho, and V. Cuperman, "Combined speech and channel coding for mobile radio," in *Proceedings of the 1992 IEEE International Conference on Communications*, 1992, pp. 180–183.

[8] K. Fazel and J. J. L'huillier, "Application of unequal error protection codes on combined source-channel coding of images," in *Proceedings of the IEEE*

International Conference on Communications, Georgia, USA, April, 1990, 1990, pp. 898–903.

[9] Y. Furuya, H.Fukagawa, and H. Matsui, "High speed mobile facsimile with error protection," in Proceedings of the 37th IEEE Vehicular Technology Conference, Florida, USA, 1–3 June, 1987, 1987, pp. 32–37.

[10] S. Ito, T. Miki, and A. Adachi, "Facsimile signal transmission in digital mobile radio," in Proceedings of the 38th IEEE Vehicular Technology Conference, Philadelphia, USA, 15–17 June, 1988, 1988, pp. 32–37.

[11] R. L. Brewster and R. S. Jalal, "Performance study of the transmission of image data to mobile terminals," in Proceedings of the 5th International Conference on Mobile Radio and Personal Communications, April, 1989, 1989, pp. 122–126.

[12] A. Heron and N. MacDonald, "Video transmission over a radio link using H.261 and DECT," in Proceedings of the International Conference on Image Processing and its Applications, 1992, 1992, pp. 621–624.

[13] H. Ibaraki, T. Fujimoto, and S. Nakano, "Mobile video communications techniques and services," in Proceedings of Visual Communications and Image Processing '95, Taipei, Taiwan, 24–26 May, 1995, 1995, pp. 1024–1033.

[14] J. B. Cain, G. C. Clark, and J. M. Geist, "Punctured convolutional codes of rate $(n-1)/n$ and simplified maximum likelihood decoding," IEEE Transactions on Information Theory, vol. IT-25, pp. 97–100, January 1979.

[15] J. Hagenauer, "Rate-compatible punctured convolutional codes (RCPC codes) and their applications," IEEE Transactions on Communcations, vol. 36, pp. 389–400, April 1988.

[16] R. M. Pelz, "An unequal error protected $p \times 8$ kbit/s video transmission for DECT," in Proceedings of the 44th IEEE Vehicular Technology Conference, 1994, 1994, pp. 1020–1024.

[17] K. Ramchandran, A. Ortega, K. M. Uz, and M. Vitterli, "Multiresolution broadcast for digital HDTV using joint source/channel coding," IEEE Journal on Selected Areas in Communications, vol. 11, no. 1, pp. 6–23, January 1993.

[18] L. Hanzo and J. Streit, "Adaptive low-rate wireless videophone schemes," IEEE Transactions on Circuits and Systems for Video Technology, vol. 5, no. 4, pp. 305–318, August 1995.

[19] J. Streit and L. Hanzo, "An adaptive discrete cosine transformed videophone communicator for mobile applications," in Proceedings of the International Conference on Acoustics, Speech, and Signal Processing, 1995, 1995, pp. 2735–2738.

[20] M. Khansari, A. Jalali, E. Dubois, and P. Mermelstein, "Low bit-rate video transmission over fading channels for wireless microcellular systems," IEEE

Transactions on Circuits and Systems for Video Technology, vol. 6, no. 1, pp. 1–11, February 1996.

[21] A. J. Viterbi, "Error bounds for convolutional codes and an asymptotically optimum decoding algorithm," *IEEE Transactions on Information Theory*, vol. IT-13, pp. 260–269, April 1967.

[22] J. G. Proakis, *Digital Communications*, McGraw-Hill, NY, 2nd edition, 1989.

[23] K. Illgner and D. Lappe, "Mobile multimedia communications in a universal telecommunications network," in *Proceedings of Visual Communications and Image Processing '95, Taipei, Taiwan, 24–26 May, 1995*, May 1995, pp. 1034–1043.

[24] R. Stedman, H. Gharavi, L. Hanzo, and R. Steele, "Transmission of subband-coded images via mobile channels," *IEEE Transactions on Circuits and Systems for Video Technology*, vol. 3, no. 1, pp. 15–27, February 1993.

[25] C. Y. Lee, *Mobile Communications Engineering*, McGraw-Hill, NY, 1982.

[26] J. D. Parsons, *The Mobile Radio Propagation Channel*, Pentech Press, London, 1992.

[27] J. I. Smith, "A computer generated multipath fading simulation for mobile radio," *IEEE Transactions on Vehicular Technology*, vol. VT-24, no. 3, pp. 39–40, August 1975.

[28] T. S. Rappaport, *Wireless Communications*, IEEE Press, NJ, 1996.

Index